THE DEFINITIVE GUIDE TO SUPPLY CHAIN BEST PRACTICES

THE DEFINITIVE GUIDE TO SUPPLY CHAIN BEST PRACTICES

COMPREHENSIVE LESSONS AND CASES IN EFFECTIVE SCM

Council of Supply Chain Management Professionals
and
Robert M. Frankel, Ph.D.,
University of North Florida

Vice President, Publisher: Tim Moore
Associate Publisher and Director of Marketing: Amy Neidlinger
Executive Editor: Jeanne Glasser Levine
Operations Specialist: Jodi Kemper
Cover Designer: Chuti Prasertsith
Managing Editor: Kristy Hart
Senior Project Editor: Betsy Gratner
Copy Editor: Karen Annett
Proofreader: Williams Woods Publishing
Indexer: Tim Wright
Compositor: Nonie Ratcliff
Manufacturing Buyer: Dan Uhrig

© 2014 by Council of Supply Chain Management Professionals
Published by Pearson Education, Inc.
Upper Saddle River, New Jersey 07458

Pearson offers excellent discounts on this book when ordered in quantity for bulk purchases or special sales. For more information, please contact U.S. Corporate and Government Sales, 1-800-382-3419, corpsales@pearsontechgroup.com. For sales outside the U.S., please contact International Sales at international@pearsoned.com.

Company and product names mentioned herein are the trademarks or registered trademarks of their respective owners.

All rights reserved. No part of this book may be reproduced, in any form or by any means, without permission in writing from the publisher.

Printed in the United States of America

First Printing November 2013

ISBN-10: 0-13-344875-4
ISBN-13: 978-0-13-344875-7

Pearson Education LTD.
Pearson Education Australia PTY, Limited.
Pearson Education Singapore, Pte. Ltd.
Pearson Education Asia, Ltd.
Pearson Education Canada, Ltd.
Pearson Educación de Mexico, S.A. de C.V.
Pearson Education—Japan
Pearson Education Malaysia, Pte. Ltd.

Library of Congress Control Number: 2013946943

This book is dedicated to the many noble thinkers and leaders who have contributed to the Council of Supply Chain Management and its predecessor organizations over the last five decades. Their noteworthy actions and scholarly virtues have served to define and eminently advance the fields of physical distribution, logistics, and supply chain management.

CONTENTS

Introduction 1

Part 1: Demand Management in the Supply Chain.............. 3

CASE 1   Dockomo Heavy Machinery Equipment, Ltd.:
Spare Parts Supply Chain Management (SCM).................. 5
    Introduction 5
    Background.. 6
    The Supply Chain Partners of Dockomo, Ltd. 7
    Current Situation 8
    Initial Analysis 8
    Forecasting Challenges.. 9
    Safety Stock Challenges 10
    Discussion Questions.. 10
    Endnotes 11

CASE 2   Silo Manufacturing Corporation (SMC)—Parts A and B:
Managing with Economic Order Quantity..................... 27
    SMC—Part A 27
    Discussion Questions.. 29
    SMC—Part B 30
    Discussion Questions.. 31
    Endnotes 32

CASE 3   Megamart Seasonal Demand Planning......................... 33
    Introduction 33
    Challenging Characteristics of Gas Grills 34
    Stakeholder Concerns 35
    The Emotional Nature of Grills 37
    Past Failures 37
    Relevant Details 38
    Your Task.. 39

CASE 4 Supply Uncertainty, Demand Planning, and Logistics Management at Goodwill Industries of Oklahoma. 43

Introduction 43
Oklahoma Goodwill Industries (OGI). 45
Supply Uncertainty and Demand Planning Issues at GIO. 47
Logistics Management Issues at GIO 49
Concerns for the Immediate Future. 50
Appendix. 55

Part 2: Supply Chain Network Design and Analysis 57

CASE 5 Bertelsmann China—Parts A and B: Supply Chains for Books 59

Bertelsmann China—Part A 59
    History of Bertelsmann. 60
    Bertelsmann Begins to Do Business in China. 60
    Chinese Culture and Impact on Western Businesses 61
    Industry Overview of the Chinese Book Market 61
    Direct Group China's Supply Chain 62
    Catalog Development Process 63
    Inbound Logistics 63
    Outbound Logistics. 64
    Book Arrival Situation at Shop Level. 64
    Inventory Situation at Shop Level 65
Discussion Questions. 66
Bertelsmann China—Part B 67
    Key Insights from Part A. 67
    Overview of Bertelsmann China 68
    Challenges for the Bertelsmann Supply Chain 68
    Bertelsmann's Quest for Improvement 68
    Insight into the Supply Chain Network of 21st Century. 68
    How Inbound Logistics Work at 21st Century. 68
    The Distribution Network in Detail 69
    Optimizing the Network. 69
    Need for Further Improvement 70
    How Logistics Service Providers Offer Value Added 70

　　　　　Improving Network Design 70
　　　　　A Long Night 71
　　　　Discussion Questions. 71
　　　　Endnotes 72

CASE 6　Carnival Corporation Food Supply Chain 81
　　　　Introduction 81
　　　　Carnival Corporation Overview 82
　　　　North American Cruise Industry 82
　　　　Giordano's Thought Process 83
　　　　Meeting Notes with Head Chef Rousseau 83
　　　　Typical Food Consumption. 84
　　　　Food Consumption Variables 84
　　　　Carnival Cruise Food Operating Policies. 84
　　　　The Food Procurement Process 85
　　　　Food Staff 85
　　　　Carnival's Food Supplier Base. 86
　　　　Competitive Benchmark Data. 86
　　　　Supplier Recommendations 87
　　　　　Contents of File Attachment. 88
　　　　　Attachment from L&M Meats 89
　　　　Giordano's Recommendations 89

CASE 7　DSM Manufacturing: When Network Analysis Meets
**　　　　Business Reality 95**
　　　　Introduction 95
　　　　History. 96
　　　　Project Description 96
　　　　The Meetings 97
　　　　Conclusion 101
　　　　Discussion Questions. 101

CASE 8　Kiwi Medical Devices, Ltd.: Is "Right Shoring" the
**　　　　Right Response? 109**
　　　　The Race of Life 109
　　　　Kiwi's Marathon Begins. 110
　　　　Kiwi's Sprint for Global Sales 110

    Kiwi's Cramped Manufacturing Footprint 111
    Kiwi's Race Turns Uphill 112
    Choosing Where to Run 113
    Running with Risk, but How Much? 114
    Selecting a Race Support Team 116
    Time to Relax—For a Moment 117
    Discussion Questions 118

Part 3: Risk and Uncertainty in the Supply Chain 123

CASE 9   Innovative Distribution Company: A Total Cost Approach to Understanding Supply Chain Risk 125

    Introduction 125
    Learning Objective and Appropriate Audience 126
    New Product Sourcing Details 127
    Domestic Supplier Details 127
    Global Supplier Details 128
    Discussion Questions 128
    Endnotes 130

CASE 10   Humanitarian Logistics: Getting Donated Foods from Switzerland to Zambia 133

    Case Overview/Background 133
    Routing of Shipment 134
    Shipment Details 135
    Transportation Documentation 136
    Receipt of Goods in Lusaka 137
    Transportation from Customs to the House of Moses 137
    Allocation of Foods to Other Aid Organizations 139
    Goods Turnover Program 140
    Repositioning of Container II 141
    Concluding Activities of the Logistics Project 141

Part 4: The Functions 147

CASE 11 Breaking Ground in Services Purchasing 149
Background.. 149
Purchasing Reorganization 150
Category Management.. 151
Legal Spending 153
Meeting with Legal Department................................... 154
Discussion Questions... 154
Endnotes 155

CASE 12 Lean at Kramer Sports................................. 165
Kramer Sports.. 165
The Employee-Owned Cooperative 166
Factory Operations at Kramer 166
Tough Times 167
Lean Implementation at Kramer 168
One-Year Update 169
Wilcox's List 170

CASE 13 UPS Logistics and to Move Toward 4PL—or Not? 173
Introduction 173
Industry Environment and Business Model Context 173
UPS Logistics Approach to 4PL 175
Managing the Change Process 177
Changing Economics and Coordination 178
Sample Client Relationship: Cisco Systems 179
Discussion Questions... 180
Endnotes 180

Index 187

ACKNOWLEDGMENTS

This book is the result of a collective effort by many leading scholars over the past decade. The Council is appreciative of the efforts of Jeanne Glasser Levine and her outstanding team at Pearson for their advice during the publication process; of Robert Frankel and Leanna Payne of the University of North Florida, whose respective editorial and administrative efforts made publication possible; and of the many academic contributors who wrote, submitted, and reviewed the case studies included herein, which serve to define the field and profession.

ABOUT THE EDITOR

Robert Frankel, Ph.D., is the Richard deRaismes Kip Professor of Marketing and Logistics at the University of North Florida. A Fulbright Scholar, he received his Ph.D. in Marketing and Logistics from Michigan State University. His research has been published in many high-profile academic journals worldwide, focusing on supply chain management, international marketing, and pedagogy.

ABOUT THE AUTHOR

Founded in 1963, the **Council of Supply Chain Management Professionals** (CSCMP) is the preeminent worldwide professional association dedicated to the advancement and dissemination of research and knowledge on supply chain management. With more than 8,500 members representing nearly all industry sectors, government, and academia from 67 countries, CSCMP members are the leading practitioners and authorities in the fields of logistics and supply chain management. The organization is led by an elected group of global officers and is headquartered in Lombard, Illinois, USA.

INTRODUCTION

The discipline and practice of supply chain management (or SCM as it is frequently known) continue to rapidly evolve. Such evolution challenges current and future managers to keep up with trends, new analytical techniques, and world-class best practices. In truth, most managers struggle with that challenge, and constantly request senior-level management to provide them with realistic scenarios wherein they can learn and improve their skill set.

SCM spans a wide array of functional activities ranging from business forecasting, supply management, and demand planning (which occur well in advance of business-customer interactions) to order fulfillment and post-sales service (which happen as or after interactions are complete). Those companies that are best at SCM—those that truly adopt SCM best practice as a driving philosophy—are the ones that can seamlessly and successfully integrate these functional activities into a set of core processes that link the company with its suppliers and customers. For these leading-edge companies, success is defined in multiple ways, including the provision of topflight customer service, the minimization of total landed costs, the balance of supply and demand, and/or the ability to continually innovate in ways that customers value, to name but a few possibilities.

Regardless of which valuable outcome(s) a company is focused on, it can unequivocally be stated that the managers of the present and future are/will be challenged to achieve such valued outcomes due to organizational design constraints. Those constraints exist both (1) *within* the organization as well as (2) *across/between* the organization and its key suppliers and its key customers. Historically, companies have been organized into functional areas, each of which specializes in certain predefined business tasks, with optimization occurring internal to the functional unit. This arrangement has allowed employee groups to become excellent at what they do best and deliver many different types of value to suppliers and customers. However, the focus on functional perfection embraced by many companies has its drawbacks, and a primary issue many otherwise effective companies face is how to create seamless interconnections between functional subgroups that possess disparate goals, value systems, and methodologies—yet are working to serve the needs of the same customer or supplier. In fact, many SCM experts believe that a company must first achieve such integration *within* the organization before it will be able to address the challenge of integrating *across/between* its key customers and key suppliers. Managers at all levels of leading organizations frequently comment that they know of very few ways to address these two integration challenges, despite their having become a perennial focus of senior executives in recent years.

For students of SCM—be they in the university classroom, the office complex, or the corporate boardroom—to effectively address the supply chain issues that will undoubtedly continue to perplex businesspeople for years to come, it is necessary for them to gain experience with a number of tasks. At a minimum, it would seem helpful for students to be tasked with (1) recognizing the types of supply chain problems that companies face, (2) gathering potential solutions, (3) evaluating plausible options, and (4) designing methods of implementation as well as measurement. "Practice makes perfect," or so the old saying goes. To this end, teachers and students alike have long availed themselves of business case studies as a viable substitute for (or a partner with) hands-on, sleeves-rolled-up experience. The benefits of case studies have been demonstrated to be numerous. Such benefits include the development of qualitative and quantitative analytic and problem-solving skills, application of new knowledge, exploration of solutions for complex issues, decision-making skills, oral and written communication skills, time management skills, interpersonal/social skills, and creativity and self-confidence.

Given the above, the purpose of this book is to provide multiple, realistic, fact-based scenarios for both academic and managerial analysis of SCM problems. We believe that students and managers of all ages and experience levels will benefit from this collection of "virtual experiences" within many different contexts that vex managers on a daily basis. The cases that make up this book are a compendium of award-winning, custom-designed scenarios written by experts commissioned by the Council of Supply Chain Management Professionals. Each case challenges the reader to consider and then address a particular issue, which reflects the complexities of modern SCM in fictitious settings based on real-world events. Sometimes, to borrow another cliché, the "truth is stranger than fiction," and the deep examination of the truths laid out within this collection should yield wisdom beyond that which would be expected from rote study of texts and articles alone.

The book is divided into four parts, each of which is concerned with a focal topic of contemporary interest and challenge in SCM. Each part contains two to four cases. The four parts are titled as follows: (1) "Demand Management in the Supply Chain," (2) "Supply Chain Network Design and Analysis," (3) "Risk and Uncertainty in the Supply Chain," and (4) "The Functions." A number of the cases within the four parts are global in nature and scope, which further enhances the value and realism of the learning scenario as the student and/or manager considers and evaluates the situations posed in search of advisable responses. We believe that you will find the cases to be an invaluable learning tool for your teaching environment.

Part 1

DEMAND MANAGEMENT IN THE SUPPLY CHAIN

This section contains four cases: (1) "Dockomo Heavy Machinery Equipment, Ltd.: Spare Parts Supply Chain Management (SCM)," (2) "Silo Manufacturing Corporation (SMC)—Parts A and B: Managing with Economic Order Quantity," (3) "Megamart Seasonal Demand Planning," and (4) "Supply Uncertainty, Demand Planning, and Logistics Management at Goodwill Industries of Oklahoma." Although the focus of the four cases is best described as demand management, each case examines a particular aspect of the demand management process, which particularly challenges managers.

The "Dockomo Heavy Machinery Equipment, Ltd.," case, which is based in India, focuses on spare parts SCM, which is a topic that receives far less attention than does the typical topic of "new" finished goods SCM. Spare parts are particularly challenging in a global environment, when supply chains cover long distances and resupply quantities are often small. Dockomo demonstrates the challenges in forecasting real demand patterns and requires students to recommend appropriate forecasting techniques and plans for implementation. Dockomo also requires determination of safety stock levels; moreover, it is concerned with demand that is not normally distributed (reflecting the real world).

The "Silo Manufacturing Corporation (SMC)—Parts A and B" case focuses on economic order quantity (EOQ) management, and is a two-part case wherein the individual parts can be utilized separately or in combination. Silo demonstrates that by introducing real-life constraints and considerations to the EOQ model, student learning is significantly enhanced. Silo considers competing performance metrics between different siloed functions in a company, the total cost concept and inventory management, the impact of changing key variables of EOQ, EOQ robustness, and just-in-time implementation.

The "Megamart Seasonal Demand Planning" case focuses on the impact of seasonality with respect to demand planning. This is a common problem that many retailers face when dealing with bulky, seasonal, or imported product categories. By allowing students to "get their hands dirty" analyzing real-world point-of-sale (POS) data, Megamart demonstrates the challenges of making very tactical decisions that then create multiple strategic implications. The case examines different functional perspectives within a company and the importance of trade-off analysis.

The "Supply Uncertainty, Demand Planning, and Logistics Management at Goodwill Industries of Oklahoma" case focuses on demand-planning decision making in a not-for-profit company. In particular, supply-and-demand balancing in this business environment is characterized by extreme supply-side uncertainty or variance. The case considers multiple inventory considerations as well as seasonality; the case also offers the opportunity to address the company's truck routing problem.

1

DOCKOMO HEAVY MACHINERY EQUIPMENT, LTD.: SPARE PARTS SUPPLY CHAIN MANAGEMENT (SCM)

A. Narayanan, University of Houston
S. Seshadri, Texas A&M University

Introduction

Only when the custodial staff showed up at his office door did Vinod Mehra realize that it was already 3:00 a.m. Vinod is the vice president of supply chain for Dockomo Heavy Machinery Equipment, Ltd. He spent the entire night analyzing the data from the Spare Parts Division in Pune, India. It was April 15, 2010, and he had just two weeks to go before the annual company review.

The Spare Parts Division's growth at Dockomo had slowed down to about 10 percent annually when compared with the growth rate of 20 percent annually over the previous years. Canceled orders stood at a staggering 8 percent due to parts unavailability, but, at the same time, the inventory in the system was $6 million higher than the previous year. Vinod was unsure of the response he would receive from the board of directors because the inventory level increased along with the number of canceled orders.

At the meeting, the board was considerate, but Vinod was asked to conduct an analysis of the shortcomings and prepare a report on the leading causes for the unavailability of parts to the customers. He was also asked to prepare a report on the approach to be followed to fix these problems by the next quarterly meeting. Vinod was already aware of many issues that existed in the supply chain, but he had to go through a complete analysis to gain a clearer understanding of the shortcomings in the distribution processes.

Background

Dockomo Heavy Machinery Equipment, Ltd. (DHEL), was founded in 1961, and at that time primarily focused on construction projects. Dockomo witnessed steady growth and its revenue reached $7 million in 1970, the same year it also became a public company. With its revenue continuously increasing in India during the 1980s and 1990s, Dockomo diversified into other industry segments, such as the technology and automobile sectors. In 1999, the net revenue from all the businesses was $10 billion. Even though it enjoyed a monopoly and high profit margin in the parts distribution market, it was still a relatively small division of Dockomo in the 1990s. During the year 2000, India witnessed many changes in public policy and the service contracts that were vital to the business growth at Dockomo. The government began to increase access to mining sites in the country, to satisfy the increasing need for energy projects. This directly impacted the heavy construction equipment industry in the country. The changes in policy meant that the demand for leased equipment would also increase dramatically.

Dockomo was proactive and understood the potential of this change in industry dynamics. The volume of the heavy machinery construction equipment in the country, which until 1999 was only 500 machines, was forecast to increase twentyfold in the next few years. The Spare Parts Division of Dockomo had to be revamped to ensure improved parts availability. Dockomo implemented an SAP system in 2000, an investment of about $5 million, which equipped the Spare Parts Division with inventory visibility and material requirements planning.

As a part of the supply chain revamp of 1999, a central distribution center (CDC) with 100,000 square feet of warehouse space was set up in Pune, India. In addition, three regional centers with 30,000 square feet of warehouse space were established at Chennai, Tamil Nadu; Kolkata, West Bengal; and Noida, Uttar Pradesh. A supply chain organization was created, with Vice President Vinod directly reporting to the board. The supply chain organization was further divided into the Procurement Division and Warehouse Operations Division. The Procurement Division consisted of a procurement head along with two procurement analysts. The Operations Department consisted of a warehouse operations head along with two inbound logistics managers and two outbound logistics managers. The functional operations within the distribution center (DC) were outsourced to VST Logistics.

Since 1999, the Spare Parts revenue grew rapidly and reached $370 million by 2010 (Exhibit 1). However, Dockomo never updated its inventory or buying policies, which were based on its 1999 environment.

The Supply Chain Partners of Dockomo, Ltd.

In 2010, three vendors were delivering parts to the CDC in Pune. Dockomo's main heavy machinery product categories were cranes (a machine for maneuvering heavy weights via a projecting swinging arm), dumpers (a vehicle designed for carrying bulk material at construction sites), and graders (a machine with a long blade used to create a flat surface). The parts master list grew from 20,000 parts in 1999 to 50,000 parts in 2010 with the addition of six new dumpers, four new graders, and six new crane models. Dockomo's partner in South Korea, EJK, was the sole vendor for its critical machines and hydraulic parts for the popular crane models (EX300, EX270, EX150, and EX70—which are 300-, 270-, 150-, and 70-ton equipment, respectively). The other vendor, KPLI in Pune, was responsible for supplying the bulk of functional parts and served Dockomo by setting up its warehouse right opposite to the CDC in Pune. The third vendor, Scan, Ltd., based out of Bangalore, Karnataka, provided some parts for the new dumper and grader models. The percentage of business served by EJK was 15 percent, KPLI was 80 percent, and Scan, Ltd., was 5 percent, as shown in Exhibit 2.

Each vendor had different delivery policy requirements. KPLI required a two-week notice, while Scan required a one-month notification. On the other hand, EJK, the international vendor, required a minimum of a three-month notification on all its parts. EJK had two different divisions, one being the Hydraulic Parts Division (HLW), which required a four-month notification, and the Mechanical Parts Division (MLW), which required a three-month notification. Emergency orders could be placed with any of these vendors, but they were assessed a 7 percent upcharge on the expedited parts ordered due to the cost of air shipment.

From Exhibits 1 and 3, it is evident that Dockomo has reached a crossroads in its spare parts business. The business volume has increased about eight times since 1999, but the warehouse space in Pune has grown only by 25 percent. The total part numbers in distribution has also more than doubled in the last 10 years. Even though Dockomo still held the majority of market share in India, there were competitors that were catching up by promising better customer service. Market survey results showed that customers preferred Dockomo as the oldest player in this business segment, with the "trust" factor being so important in India. At the same time, customers voiced their opinions over dissatisfaction with parts availability. About 30 percent of the customers surveyed said that if they were given an option of opting out of the contract without penalty, they would shift to another heavy equipment provider. This dissatisfaction was a cause of concern for Dockomo.

Before starting the review, Vinod had collected data about other equipment manufacturers in India. Exhibit 4 shows the market share of Dockomo and its competitors. As expected, Dockomo still has the largest market share of 52 percent followed by Butterfly, Inc., at 24 percent. But, Butterfly's growth has been startling, with 250 percent growth in the last four years while Dockomo could only manage a growth of 110 percent during the same time period. There is also an emergence of a gray market, which threatened the availability of genuine parts. Sold by small businesses in mining areas with a high density of construction equipment, these parts are similar to the genuine parts, but sold at a much cheaper rate. Currently, no corporate or legal action is under way to control this gray market; the only visible effort is the penalty on warranty and service contracts if a gray-market product is used by the customer.

Current Situation

The supply chain organization reduced its personnel from nine people in 1999 to seven people in 2010. There are only two procurement planning analysts and each person is responsible for planning about 50,000 parts. In 2010, they recorded a forecasting error of 70 percent for the top 5 percent of items, which contribute to the majority of their revenue and profit. The first pick rate has also dropped drastically. *First pick rate* refers to the availability of parts in the warehouse when the customer requests them. The first pick rate of Dockomo was a dismal 29 percent in April 2010, compared with 48 percent in April 2008.

Dockomo hired a supply chain consulting firm, Thomas Consulting, to analyze the shortcomings in its distribution network. The project manager and senior analyst at Thomas Consulting visited all the DCs to understand the procurement process. Their analysis showed that the order policy at Dockomo is based on average monthly consumption over the last one year. The average is a very poor indicator of the consumption in the spare parts business. It was very clear that Dockomo needed a more proactive system where the demand patterns, rather than the consumption patterns, would be considered. To counter this uncertain demand, Dockomo carried a fair amount of safety stock. The amount of safety stock carried at any DCs was calculated based on the buyer's experience: anywhere between 1.2 to 2 times the average demand, depending on the volatility of the demand and lead time of the product. The part categorization was based on the cost of the part. Lower cost products were procured every three or six months to achieve some economies of scale in procurement costs. There was also mistrust between the zonal DCs and the CDC, which led to panic ordering and a lack of data integrity in the system.

Initial Analysis

The analysts first stratified the inventory based on ordering frequency to attain some visibility. The analysis revealed that Dockomo was carrying about $30 million in inventory all over India, with about one third being classified as dead—parts that have not been ordered by the customer over the last two years. Another $7 million in inventory was engaged in parts that have been ordered less than five times over the previous year. The overall average annual inventory turn is about 2.4, well below the industry average of 4.0 for this industry in India. Parts were categorized into fast movers, medium movers, slow movers, very slow movers, and dead inventory, as shown in Table 1. Dockomo's CDC was carrying $10 million worth of inventory at the time of analysis. With the arrival of new part designs, the spare parts of the older products did not move and became stagnant. Dockomo was planning to phase out this dead inventory over the next year.

To design the solution, Vinod provided data of the top 20 percent of SKUs in each category. The consultants had to devise a strategy to design forecasting and safety stock techniques for the given SKUs[1] (Tables 2 and 3a–3d).

Forecasting Challenges

Vinod and the consultants agreed that a proactive supply chain with systematic forecasting could solve the issue of the parts availability, but, currently, forecasting is based on a planner's experience rather than mathematical forecasting techniques. Analysis of the data showed that very few SKUs followed time-series characteristics, such as trends or seasonality. Out of the SKUs given, about 20 percent had trend characteristics, another 20 percent showed seasonality traits, while the rest had erratic and lumpy demand. For example, item Cr106 has a positive trend, while Cr101 shows patterns of seasonality with peaks at the beginning of the year and valleys midway through the year. But many SKUs are like Cr102 and show no definite patterns, making it difficult to predict (see Exhibit 5). At present, the planners try to forecast almost all the items at least once every year and they do it as stated earlier, based on their experience. For the fast and medium movers, they update the forecast at regular intervals, either monthly or quarterly. Because these items have different demand patterns, there is a need to first identify the items that can be forecasted and then to pair them with the appropriate forecasting technique.

Another challenge in forecasting is the vendor in Korea, EJK, which requires a three-month notification for mechanical parts and a four-month notification for hydraulic parts. It is very difficult to forecast three months ahead, especially with the growing business and unpredictable demand patterns. EJK cannot relax its lead-time requirement as it is constrained by the lead time of securing raw materials to manufacture the part. Vinod and the consultants face the challenge of designing a forecasting framework that would account for all demand patterns, while taking into account the vendor lead-time requirements.

Safety Stock Challenges

The biggest cause of concern for the group was the increasing stockouts. Dockomo had no tracking of stockouts in its inventory system—the staff just knew that they had too many customer complaints about product unavailability over the last year. The consultants had to compute the approximate number of stockouts for each item based on order, inventory, and DCs' shipment data for the last six months. Exhibit 6 shows the stockout percentage by item type. This figure was a pleasant surprise for the team, as percentage of stockouts was relatively constant for each category. There were about 8 percent for fast movers, 10 percent for medium movers, 3 percent for slow movers, and 2 percent for very slow movers. As explained earlier, Dockomo uses no statistical procedure to compute its safety stock; it is simply a multiple of average consumption.

Calculation of safety stock in most enterprise resource planning (ERP) packages is based on the normal distribution (i.e., assuming that the underlying demand is normally distributed). But, on analyzing the demand patterns, the team found that less than 3 percent of the SKUs were normally distributed.[2] Most of the distributions were skewed (extremely right-tailed). For example, Exhibit 7 shows demand distributions of SKUs across three categories. Items Cr108 and Cr137 are skewed to the left, whereas Cr106 is approximately normally distributed. These demand patterns present a unique challenge to the team because if the normal distribution is assumed, it results in incorrect safety stock levels. The safety stock has to be computed using an alternative method of accounting for this highly uncertain demand pattern. Dockomo wants to promise a service level of at least 95 percent (i.e., order fill rate of 95 percent)[3] to all customers and achieve a first pick rate of 70 percent by the end of this implementation.

Vinod and his consultant team have exactly two months to come up with a strategy to fix Dockomo's forecasting and safety stock issues.

Discussion Questions

1. How can the conundrum of erratic forecasts be solved? Should Vinod consider forecasting all the parts in the SKU master list (F, M, S, VS) or only a subset of categories?
2. How should the safety stock problem be approached? If they don't satisfy normal distribution conditions, what methodology should be used?
3. What should be done to accommodate the parts from vendor EJK, which has lead times of three months (machine parts) and four months (hydraulic parts)?
4. How would you estimate the reorder point of all parts and what should be the item fill rate across the product categories to achieve a service level of 95 percent?

Endnotes

1. Due to space constraints, data has been provided for the top 10 SKUs in each category (F, M, S, VS). Data for the Z category has not been provided as they would be phased out of the inventory.
2. Kolmogorov Smirnov Goodness of fit test was used to verify normal distribution.
3. There are items with more than 95 percent fill rate, like the slow and very slow movers.

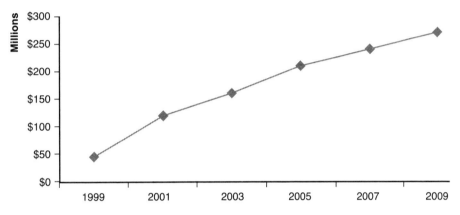

Exhibit 1   Yearly revenue of Spare Parts Division (1999 to 2009)

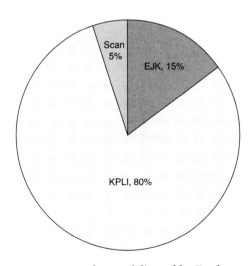

Exhibit 2   Percentage of parts delivered by Dockomo vendors

Case 1 Dockomo Heavy Machinery Equipment, Ltd.   11

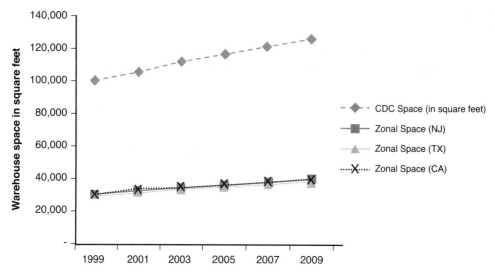

Exhibit 3   Growth of distribution centers (1999 to 2009)

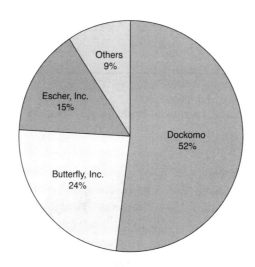

Exhibit 4   Heavy equipment market share in India

Table 1 Dockomo Parts Categorization Proposed by Thomas Consulting

Order Frequency per Year	Categories	Total Number of SKUs	Percentage
Greater than 200	Fast Movers (F)	450	0.90%
Between 50 and 200	Medium Movers (M)	1,250	2.50%
Between 10 and 50	Slow Category (S)	12,000	24.00%
Less than 10	Very Slow Category (VS)	18,000	36.00%
Parts not ordered in the last two years	Dead (Z)	18,300	36.60%

Table 2 Inventory List with Corresponding Category, Vendor, Price, and Lead Time

SKU	Category	Vendor	Price ($)	Lead Time (Months)
Cr101	Fast Movers	EJK (HLW)	1,987	4
Cr102	Fast Movers	Scan	161	1
Cr103	Fast Movers	EJK (MLW)	234	3
Cr104	Fast Movers	EJK (MLW)	1,944	3
Cr105	Fast Movers	KPLI	1,725	1
Cr106	Fast Movers	KPLI	892	1
Cr107	Fast Movers	KPLI	1,987	1
Cr108	Fast Movers	Scan	1,898	1
Cr109	Fast Movers	KPLI	462	1
Cr110	Fast Movers	KPLI	1,098	1
Cr111	Medium Movers	EJK (MLW)	433	3
Cr112	Medium Movers	KPLI	2,801	1
Cr116	Medium Movers	KPLI	127	1
Cr118	Slow Movers	KPLI	3,062	1
Cr121	Very Slow Movers	EJK (MLW)	3,211	3
Cr122	Medium Movers	KPLI	2,388	1
Cr126	Very Slow Movers	KPLI	9,921	1
Cr131	Slow Movers	KPLI	7,796	1
Cr132	Medium Movers	KPLI	1,732	1
Cr133	Slow Movers	KPLI	4,883	1
Cr137	Very Slow Movers	EJK (HLW)	49,121	4

Table 2 Continued

SKU	Category	Vendor	Price ($)	Lead Time (Months)
Cr166	Slow Movers	EJK (MLW)	8,769	3
Cr166	Very Slow Movers	KPLI	7,771	1
Cr222	Slow Movers	KPLI	22,844	1
Cr321	Medium Movers	EJK (MLW)	4,208	3
Cr343	Very Slow Movers	KPLI	4,323	1
Cr573	Very Slow Movers	KPLI	6,543	1
Cr645	Slow Movers	KPLI	3,546	1
DM102	Medium Movers	EJK (HLW)	3,321	4
DM109	Very Slow Movers	KPLI	1,322	1
DM115	Very Slow Movers	Scan	5,023	1
DM124	Medium Movers	KPLI	1,517	1
DM234	Medium Movers	KPLI	1,656	1
DM245	Slow Movers	KPLI	10,176	1
DM333	Medium Movers	KPLI	3,178	1
Gr100	Very Slow Movers	KPLI	5,437	1
Gr114	Slow Movers	KPLI	11,936	1
Gr122	Slow Movers	EJK (HLW)	3,189	4
Gr355	Very Slow Movers	KPLI	71,200	1
Gr365	Slow Movers	KPLI	13,349	1

Due to space constraints, data has been provided for the top 10 SKUs in each category (F, M, S, VS). Data for the Z category has not been provided as they will be phased out of the inventory.

Table 3a Monthly Demand Data for Fast Movers

Month	Cr101	Cr102	Cr103	Cr104	Cr105	Cr106	Cr107	Cr108	Cr109	Cr110
Jan '07	306	11	3	70	38	24	181	70	14	100
Feb '07	367	90	14	98	9	41	93	98	22	200
Mar '07	294	2	0	71	28	45	60	71	4	255
Apr '07	286	0	0	56	3	56	3	47	9	265
May '07	298	90	70	48	12	48	67	55	33	0
June '07	322	10	15	47	25	49	46	30	69	5
July '07	357	34	22	66	0	48	34	70	0	0
Aug '07	380	46	18	93	14	42	67	93	9	1
Sept '07	484	20	0	107	6	39	28	107	3	65
Oct '07	512	64	14	116	26	56	87	148	19	5
Nov '07	544	78	11	121	21	50	169	133	9	142
Dec '07	503	120	0	124	1	66	29	108	0	10
Jan '08	497	99	0	126	17	74	92	112	5	0
Feb '08	454	162	0	133	10	77	83	79	15	2
Mar '08	390	338	98	118	6	68	50	67	2	31
Apr '08	388	27	14	112	8	66	0	78	0	32
May '08	422	39	1	108	0	78	27	70	7	79
June '08	434	26	0	102	8	82	38	60	0	36
July '08	456	54	0	108	0	71	0	50	6	169
Aug '08	484	160	22	114	0	86	0	114	1	260
Sept '08	501	145	32	128	2	90	95	78	0	501

Table 3a Continued

Month	Cr101	Cr102	Cr103	Cr104	Cr105	Cr106	Cr107	Cr108	Cr109	Cr110
Oct '08	544	553	1	133	0	96	108	102	17	246
Nov '08	480	149	34	140	5	103	125	114	6	416
Dec '08	453	75	14	144	0	110	111	128	45	399
Jan '09	454	23	62	148	14	119	0	120	7	212
Feb '09	460	11	0	156	0	127	101	113	20	303
Mar '09	403	6	7	144	2	133	89	112	3	590
Apr '09	392	0	0	139	0	141	61	83	12	177
May '09	403	11	56	141	0	166	195	86	9	949
June '09	434	6	61	132	1	158	52	107	21	137
July '09	448	134	21	140	5	168	53	109	3	214
Aug '09	452	148	14	148	3	179	112	104	132	678
Sept '09	501	185	1	156	11	181	171	111	55	100
Oct '09	492	161	13	168	8	189	163	108	17	52
Nov '09	503	259	19	172	4	192	84	132	23	165
Dec '09	484	80	30	176	33	196	68	140	31	95
No. of Orders (Order Lines Data [O.L]) in the Last Three Years ('07, '08, '09)										
O.L ('07)	204	284	257	255	205	202	289	286	290	345
O.L ('08)	234	216	216	311	224	207	278	250	242	262
O.L ('09)	245	312	284	291	241	305	306	344	233	300

Table 3b Monthly Demand Data for Medium Movers

Month/SKU	Cr111	Cr112	DM124	DM333	DM234	Cr321	Cr122	DM102	Cr116	Cr132
Jan '07	103	6	1	5	225	61	10	79	4	81
Feb '07	0	16	19	4	413	52	11	20	6	100
Mar '07	22	1	4	4	147	67	9	59	8	113
Apr '07	0	4	24	65	96	61	14	31	12	125
May '07	14	6	18	50	253	68	7	39	14	139
June '07	1	12	32	6	220	75	11	27	2	124
July '07	3	0	43	236	242	80	7	34	0	129
Aug '07	0	4	23	0	419	81	12	11	2	129
Sept '07	4	4	9	55	415	77	7	12	1	127
Oct '07	18	7	12	52	158	70	9	61	5	114
Nov '07	2	13	27	0	0	78	9	21	0	100
Dec '07	16	0	6	246	15	61	6	14	0	101
Jan '08	10	8	2	6	1241	61	6	14	2	101
Feb '08	16	1	4	0	715	56	14	18	0	108
Mar '08	2	5	4	20	749	47	12	67	1	107
Apr '08	0	0	29	6	204	44	7	61	2	91
May '08	6	0	2	13	877	41	14	2	0	79
June '08	4	0	0	4	509	42	14	13	20	85
July '08	4	1	0	10	244	41	13	2	31	78
Aug '08	2	7	0	3	1045	39	8	3	2	73
Sept '08	38	2	8	0	235	40	7	6	4	77

Table 3b Continued

Month/SKU	Cr111	Cr112	DM124	DM333	DM234	Cr321	Cr122	DM102	Cr116	Cr132
Oct '08	27	2	2	11	436	42	10	25	2	90
Nov '08	39	0	16	10	273	47	11	6	0	100
Dec '08	23	2	2	2	605	53	6	9	27	108
Jan '09	29	1	0	5	474	56	8	7	0	129
Feb '09	10	5	16	12	451	59	7	1	27	120
Mar '09	0	1	8	40	697	68	8	0	27	124
Apr '09	15	10	9	11	481	73	6	3	33	125
May '09	13	0	7	7	519	83	6	23	56	131
June '09	3	5	3	15	440	88	12	2	0	140
July '09	43	4	13	20	312	87	12	8	20	151
Aug '09	0	8	14	25	511	90	15	16	56	164
Sept '09	0	1	37	8	570	83	11	4	33	163
Oct '09	12	4	2	16	399	98	7	26	51	170
Nov '09	22	14	10	54	655	95	7	32	88	163
Dec '09	0	8	7	0	504	81	12	36	102	148

No. of Orders (Order Lines Data [O.L]) in the Last Three Years ('07, '08, '09)

O.L ('07)	79	73	134	74	145	66	112	113	79	73
O.L ('08)	100	69	69	85	148	111	87	143	100	69
O.L ('09)	96	61	92	58	127	139	59	98	96	61

Table 3c Monthly Demand Data for Slow Movers

Month	Cr131	Cr133	Cr118	Cr166	DM245	Gr122	Gr365	Gr114	Cr645	Cr222
Jan '07	5	52	3	0	12	1	5	1	4	3
Feb '07	4	0	0	3	12	10	0	24	6	0
Mar '07	1	25	4	4	6	1	0	0	0	4
Apr '07	1	0	25	5	0	0	7	0	0	0
May '07	0	0	23	0	0	2	14	0	0	0
June '07	2	26	36	0	6	1	20	0	0	4
July '07	3	0	15	6	0	0	13	22	4	0
Aug '07	6	0	4	0	0	1	8	13	0	0
Sept '07	0	26	9	5	0	0	19	6	5	9
Oct '07	2	41	0	7	2	0	11	18	0	1
Nov '07	1	0	18	0	4	0	13	6	0	12
Dec '07	0	26	4	0	5	1	8	15	0	1
Jan '08	0	0	3	0	0	1	30	18	0	18
Feb '08	0	0	16	0	5	0	18	94	0	0
Mar '08	4	26	8	2	6	6	28	89	10	2
Apr '08	2	0	20	0	6	1	26	6	0	1
May '08	0	0	20	10	0	0	26	3	0	0
June '08	0	26	21	0	0	1	1	50	0	2
July '08	0	52	0	0	6	2	5	7	18	6
Aug '08	0	0	0	2	5	5	9	26	2	3
Sept '08	0	56	26	0	6	1	19	0	10	0

Table 3c Continued

Month	Cr131	Cr133	Cr118	Cr166	DM245	Gr122	Gr365	Gr114	Cr645	Cr222
Oct '08	0	26	2	1	7	4	27	77	12	7
Nov '08	0	52	0	2	30	0	24	57	0	0
Dec '08	0	2	34	12	0	5	28	13	12	0
Jan '09	4	178	12	1	0	2	7	0	0	8
Feb '09	0	89	7	0	4	3	8	0	0	0
Mar '09	0	26	14	1	0	3	24	56	0	0
Apr '09	0	116	5	0	14	1	24	44	0	9
May '09	5	79	10	0	6	2	20	66	0	0
June '09	2	64	10	1	0	2	16	0	0	0
July '09	0	58	0	0	45	5	19	18	55	1
Aug '09	2	0	0	5	0	4	20	29	0	6
Sept '09	1	52	5	1	0	0	24	64	10	0
Oct '09	2	0	14	0	0	13	23	62	40	7
Nov '09	6	72	0	1	18	0	21	60	25	0
Dec '09	0	0	14	0	0	2	20	92	0	9

No. of Orders (Order Lines Data [O.L]) in the Last Three Years ('07, '08, '09)

O.L (07)	12	28	19	13	20	13	32	18	13	16
O.L (08)	11	18	16	16	22	11	21	37	17	21
O.L (09)	12	21	21	10	19	14	29	48	39	11

Table 3d Monthly Demand Data of Very Slow Movers

Month	Cr121	Cr126	Cr166	Cr137	Cr343	DM115	DM109	Gr100	Gr355	Cr573
Jan '07	12	0	0	3	4	1	0	3	0	10
Feb '07	0	0	0	0	0	0	0	0	0	10
Mar '07	0	0	0	2	0	0	0	0	0	0
Apr '07	0	0	12	0	0	0	33	4	0	0
May '07	0	4	0	0	6	0	0	0	0	0
June '07	0	0	0	0	0	0	0	0	934	0
July '07	0	0	0	0	0	1	0	0	0	0
Aug '07	24	7	0	0	0	0	0	0	0	12
Sept '07	0	0	0	0	0	0	0	5	0	0
Oct '07	3	0	0	0	0	0	0	0	0	0
Nov '07	0	0	0	0	0	0	56	0	0	0
Dec '07	6	0	0	0	0	1	0	7	0	0
Jan '08	0	0	0	0	0	0	0	0	0	0
Feb '08	0	0	0	0	2	0	0	0	0	10
Mar '08	6	0	0	1	0	0	0	0	0	0
Apr '08	0	0	0	0	0	0	0	1	0	0
May '08	0	0	0	0	0	4	0	0	0	0
June '08	0	6	0	0	0	0	0	0	0	0
July '08	3	0	0	0	0	0	23	1	0	10
Aug '08	0	0	0	0	0	0	51	0	0	0
Sept '08	33	1	14	0	0	0	66	0	0	0
Oct '08	0	0	0	0	0	4	34	1	0	0

Table 3d Continued

Month	Cr121	Cr126	Cr166	Cr137	Cr343	DM115	DM109	Gr100	Gr355	Cr573
Nov '08	0	0	0	0	0	9	0	0	1080	0
Dec '08	12	0	0	0	0	4	0	0	0	0
Jan '09	0	0	16	0	0	0	0	0	0	8
Feb '09	0	0	0	1	5	0	0	0	0	0
Mar '09	7	0	0	0	0	2	0	0	556	0
Apr '09	0	0	0	0	0	0	0	2	0	317
May '09	0	2	0	2	0	0	34	0	0	4
June '09	8	0	0	4	0	1	0	0	0	192
July '09	0	0	0	0	4	0	0	5	0	10
Aug '09	0	4	2	0	0	0	0	0	0	0
Sept '09	0	0	0	1	0	0	23	0	0	0
Oct '09	0	0	0	1	4	2	0	5	0	0
Nov '09	15	6	0	0	0	0	0	0	0	0
Dec '09	0	0	0	0	5	0	21	0	0	0

No. of Orders (Order Lines Data [O.L]) in the Last Three Years ('07, '08, '09)										
O.L (07)	7	2	1	2	2	3	6	4	9	3
O.L (08)	9	2	1	1	1	4	4	3	9	2
O.L (09)	6	3	4	5	4	3	7	3	8	8

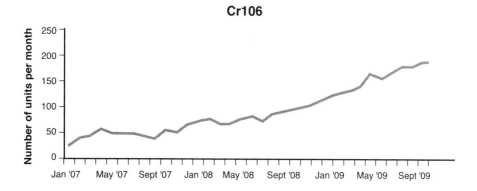

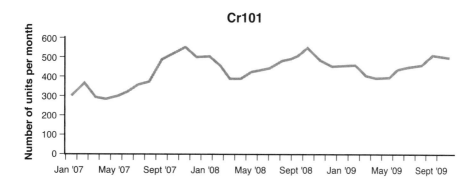

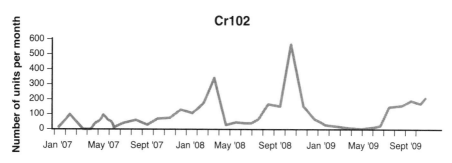

Exhibit 5   Demand patterns for Cr106, Cr101, and Cr102

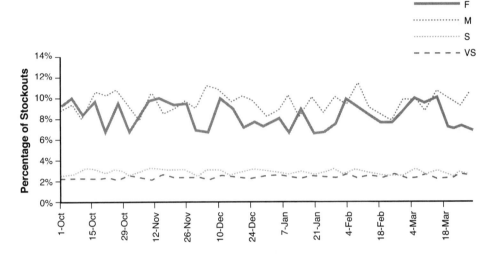

Exhibit 6 Categorization of stockouts based on Thomas Consulting recommendation (starting in October 2009)

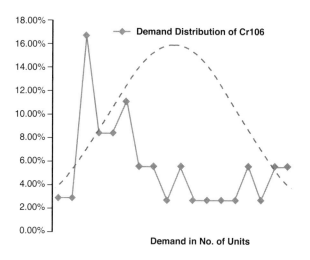

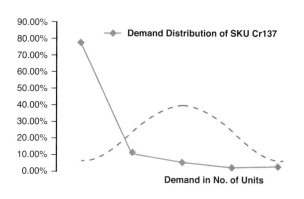

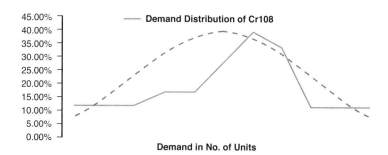

Exhibit 7 Demand distributions for Cr106, Cr137, and Cr108 with ideal normal distributions

2

SILO MANUFACTURING CORPORATION (SMC)—PARTS A AND B: MANAGING WITH ECONOMIC ORDER QUANTITY

Dr. Ted Farris, University of North Texas

SMC—Part A

SMC recently renamed itself from Silo Manufacturing Corporation to give the impression it is a modern manufacturer. Located in Blacksburg, Virginia, SMC is a regional provider of grain towers/silos for farms as far north as University Park, Pennsylvania, as far west as Knoxville, Tennessee, and as far south as Statesboro, Georgia. In recent years, SMC has extended its business to include the latest in agricultural engineering services for elevator design and installation. Its core business still remains the fabrication of the grain elevators.

Vice President of Manufacturing Ferris Martin stopped by the office of SMC's President Robert Lewin and remarked, "I need your help resolving an issue between our Financial Comptroller, Fred Ferguson, and our Purchasing Director, Peter Patrachalski. These two executives continue to argue with each other about our ordering policies." "How can I help?" asked Lewin, peering over his glasses. "Both Fred and Peter are pretty strong willed and protective of their areas."

"It boils down to conflicting goals," replied Martin. "Ferguson says the cost to carry inventory is 32 percent and is trying to keep inventory costs low. Patrachalski had his intern identify his ordering costs and was shocked to find that every time our employees place an order it costs us $48 regardless of the quantity ordered. Each one would like to dictate how the other operates so they can achieve their own performance goals. I'd like to have them meet somewhere in the middle but I'm not sure if that is the best solution."

"SMC's primary performance goals are to reduce cost and increase profitability," exclaimed Lewin. "These guys need to understand SMC comes first. Offer them a test case to propose and defend their ordering policies and we'll sort this out."

"I suggest part number 64-1909," replied Martin. "The unit cost is $112.00 and we order 10,752 units per 365-day year. Although we do not have to order by the case, it does come in 15 units per case. The average lead time from when we place the order to the time we receive it at our dock is 8.2 days with a standard variation of 1.7 days."

Later that week in the conference room, Ferguson and Patrachalski each offered proposals for ordering part number 64-1909. Purchasing Director Patrachalski stated he was trying to keep his purchasing costs down by ordering in larger quantities and suggested buying 32 cases at a time. He has also indicated he would like to avoid ordering in partial cases because doing so might result in shipments of incorrect quantities and consequent higher costs. Comptroller Ferguson claimed the most important issue was the cost to carry inventory and argued for ordering 4 cases at a time to keep average inventories low. Seeking a compromise, Lewin suggested using economic order quantity (EOQ).

Lewin stated, "Economic order quantity can be very complex. The original EOQ, known as 'Wilson's EOQ,' was actually developed by F. W. Harris[1, 2] in 1913, but a consultant named R. H. Wilson, who embraced the model and applied it extensively, was given credit for his early in-depth analysis of it.[3, 4] It determines the lowest total inventory cost by calculating the optimum order quantity denoted as Q*. Economic order quantity incorporates the trade-off between inventory carrying cost and ordering cost—exactly the trade-off we are facing with Finance and Purchasing.

"You can now find more complicated economic order quantity models extending the concept to consider discount pricing for ordering in larger quantities, back-ordering costs, differences in transportation rates if you ship by full truckload instead of LTL, including the step function of adding another warehouse as it impacts inventory carrying costs, or bridging into optimal production quantities. Anything that might influence the economic order quantity variables—there is probably an extension. There is probably even one considering the phases of the moon!

"There are a lot of assumptions for economic order quantity including the following:[5]

- A continuous, constant, and known rate of demand
- A constant and known replenishment or lead time

- Entire order delivered at same time—no in-transit inventory
- All demand is satisfied
- A constant price or cost that is independent of the order quantity (i.e., no quantity discount)
- No inventory in transit
- One item of inventory or no interaction between items
- Infinite planning horizon
- Unlimited capital

"But we should just use the original Harris-Wilson Model and consider tweaking it later. As I recall, the basic formula is:"

$$Q^* = \sqrt{\frac{2 \times \text{Annual Demand} \times \text{Ordering Cost}}{\text{Inventory Carrying Cost} \times \text{Unit Price}}}$$

At the end of the meeting, Martin agreed to take the proposals and summarize them in the following chart:

	Order Quantity (Units)	Number of Cases per Order	Orders per Year	Annual Ordering Cost	Annual Internal Cost of Capital (ICC)	Annual Total Cost
Ferguson						
EOQ to Nearest Whole Case						
EOQ						
Patrachalski						

Discussion Questions

1. What is the cost difference between Ferguson's proposal to order 4 cases each time and Patrachalski's proposal to order 32 cases each time?

2. Lewin suggested looking at economic order quantity. Based on the lowest total annual cost, what order quantity should Martin recommend?

3. Let's explore the concept of "robustness." Lewin's proposal to use economic order quantity may be unrealistic because SMC would like to place orders in

whole cases. If the order quantity is *decreased* to the nearest whole case (which is a 2.78 percent reduction), what percent would the total annual cost change? What percent would the annual total cost change if the order quantity is *increased* to the nearest whole case? *Hint:* Use the formula ([New Total Cost / Old Total Cost] − 1).

SMC—Part B

Vice President of Manufacturing Ferris Martin stopped by the office of SMC's President Robert Lewin and remarked, "Robert, your suggestion last month to use economic order quantity (EOQ) has already helped us optimize the trade-offs between inventory carrying costs and order costs."

"We are doing a better job of working as a cohesive company," replied Lewin, "but we are still faced with conflicting management goals. I recently hired Ed Davis as our new inventory control manager and plan to task him with the goal of reducing our corporate inventories by 8.9 percent at the next executive strategic roundtable. There are still many people in our company who think it is a zero-sum game and Ed will have to manage his operations carefully so he does not step on any toes."

Two weeks later, Ed Davis attended the executive strategic roundtable. The first agenda item was the assignment of annual performance goals. The 8.9 percent goal to reduce corporate inventories did not come as a surprise. In fact, Ed had already started looking at opportunities to reduce order quantities to meet the goal. Because Lewin supported the implementation of economic order quantity, Davis knew he would have to manage the variables used in EOQ in order to succeed.

During the roundtable meeting, many aggressive, yet obtainable, performance goals were proposed. Davis was pleased to see Vice President of Sales Steve Smith, Financial Comptroller Fred Ferguson, and Purchasing Director Peter Patrachalski all agree to goals to improve their areas. Davis left the meeting encouraged at the camaraderie and was delighted that he had joined such a forward-thinking organization. In this organizational culture, he might even be able to implement a just-in-time inventory system!

The next morning after a restful sleep, Davis sat down to his double espresso and orange-cranberry-pistachio biscotti to mull over his new performance goal to reduce corporate inventories by 8.9 percent. It suddenly dawned on him: Smith, Ferguson, and Patrachalski's goals all made strategic sense, but if Smith increased annual demand, the economic order quantity and corporate inventory levels would increase. Likewise, if Ferguson and Patrachalski were also successful, his inventory levels would further escalate. Their success would exacerbate his failure. The only action he could take to positively counter these "improvements" was to reduce the cost of placing an order. Getting to just-in-time may be another thing all together!

His dilemma reminded him of a quote from Albert Einstein:[6]

> *"The mere formulation of a problem is far more often essential than its solution, which may be merely a matter of mathematical or experimental skill. To raise new questions, new possibilities, to regard old problems from a new angle requires creative imagination and marks real advances."*

He would utilize the details for part number 64-1909 as a basis to determine what he needed to do to meet his inventory reduction goal. Last year, part number 64-1909 used 10,752 units at an average cost per unit of $112. The cost to place an order was $48 regardless of the quantity ordered and the SMC cost to carry inventory was 32 percent.

Discussion Questions

4. What would the cost to place an order need to be for Davis to meet his inventory reduction objective if Vice President of Sales Steve Smith achieves his goal to increase sales by 9.6 percent? (*Hint:* A 10 percent increase in sales of 100 units results in sales of 110 units.)

5. What would the cost to place an order need to be for Davis to meet his inventory reduction objective if Vice President of Sales Steve Smith achieves his goal to increase sales by 9.6 percent *and* Financial Comptroller Fred Ferguson achieves his goal of reducing the cost to carry inventory from 32.0 percent to 29.4 percent?

6. What would the cost to place an order need to be for Davis to meet his inventory reduction objective if Vice President of Sales Steve Smith achieves his goal to increase sales by 9.6 percent *and* Financial Comptroller Fred Ferguson achieves his goal of reducing the cost to carry inventory from 32.0 percent to 29.4 percent *and* Purchasing Director Peter Patrachalski achieves his goal of reducing the average cost per unit by 5.2 percent? (*Hint:* A 10 percent reduction in a $10 unit cost results in a $9 unit cost.)

7. What would the cost to place an order need to be if Davis implemented a just-in-time approach so ordering 1 unit at a time is the optimal ordering quantity? Use the original variables for part number 64-1909. Your answer must be accurate to six decimal places (e.g., $47.123456).

8. Provide three viable recommendations that would result in a lower *cost to place an order*. Keep in mind that providing a "viable recommendation" means you must move beyond theoretical statements and be immediately actionable. Do not skimp on your details (but don't make anything up either)…you must indicate *how* to lower the cost in order to receive full points.

Endnotes

1. F. W. Harris, "How Many Parts To Make At Once," *Factory, The Magazine of Management* 10, no. 2 (1913): 135–136, 152.

2. F. W. Harris, *Operations Cost*, Factory Management Series (Chicago: Shaw, 1915).

3. A. C. Hax and D. Candea, *Production and Operations Management* (Prentice-Hall, Englewood Cliffs, NJ: Prentice-Hall, 1984), 135, http://catalogue.nla.gov.au/Record/772207.

4. R. H. Wilson, "A Scientific Routine for Stock Control," *Harvard Business Review* 13 (1934): 116–128.

5. John Coyle et al., *Supply Chain Management: A Logistics Perspective*, 8th ed. (Mason, Ohio: Southwestern Cengage Learning, 2009).

6. "Albert Einstein." Quotes.net. STANDS4 LLC, 2011. 21 February 2011. http://www.quotes.net/quote/9281.

3

MEGAMART SEASONAL DEMAND PLANNING

Rodney Thomas, Ph.D., Georgia Southern University

Introduction

Megamart is a big-box home improvement retailer located in the United States. The company sells products that help customers build, enhance, and enjoy their homes. To meet the various home improvement needs of its customers, Megamart offers a complete line of products and services for decorating, maintaining, repairing, and remodeling residential buildings. With more than 2,000 retail locations, Megamart stores stock approximately 40,000 items in 26 different product categories.

Competition in the home-improvement industry is fierce. Large national chains like Lowe's and Home Depot offer name-brand products at low prices that appeal to cost-conscious consumers. Smaller regional chains often provide high levels of service and personalized attention to their customers. To remain competitive and differentiate itself from the competition, Megamart must simultaneously offer its customers low prices, wide assortments, and exceptional customer service. Therefore, efficient, effective, and responsive supply chain management is a critical area of strategic focus.

Megamart normally does an excellent job of getting the right products to the right stores at the right times in order to drive sales and meet customer demands. However, there are a few categories that are particularly challenging due to unique product and demand characteristics. Specifically, in categories that have bulky, seasonal, imported (BSI) products, Megamart struggles to meet its supply chain management goals. Therefore, CEO Theodore Esper has made the management of BSI products a top priority.

Steve Manrodtner is the new vice president of supply chain planning. Unfortunately, Mr. Manrodtner's entire career has been in transportation roles and he is not as familiar with demand management and planning principles. Therefore, he has contacted your team to develop a demand management and purchasing plan to appropriately flow BSI products to Megamart stores in a way that meets the needs of all key stakeholders. Due to the risks involved with this type of project, Mr. Manrodtner has decided that your team can conduct a BSI pilot program with stores located in one distribution center service area on a few gas grill SKUs from a single vendor. If the BSI pilot program is successful, the process you develop may be implemented throughout the retail chain.

Challenging Characteristics of Gas Grills

Gas grills are one of the most challenging assortments for supply chain managers at Megamart for several reasons. First, all grills are 100 percent imported from China and the suppliers mandate a five-month lead time for manufacturing grill orders. When international transportation and distribution lead times are added to the manufacturing lead time, the overall order cycle time for gas grills is six months. Megamart purchasing managers have tried to find alternative sources of supply, but other cost-effective options are simply not available.

The purchasing managers have also tried to negotiate shorter manufacturing lead times, but the suppliers will only operate on a build-to-order model where raw materials will not be sourced until a firm order is received from Megamart. This type of contract manufacturing approach is common in the grill industry. Therefore, assume that it takes six months for a grill to arrive in a store location once a firm order is placed. Although Mr. Manrodtner is working with other groups to reduce this overall order cycle time, consider it a firm constraint in your analysis for this project.

The demand characteristics of grills also make the assortment difficult to manage. Grill sales are highly seasonal (see Exhibits 1 and 2) and often depend on the type of weather in local markets. When it is sunny and pleasant outside, grill sales spike up. However, when it is rainy or cold, grill sales stagnate. In addition to the weather-driven seasonality of grill sales, there are also some key national holidays that create additional seasonal spikes in demand. Most stores experience large sales lifts around Memorial Day, the Fourth of July, and Labor Day. Typically, if the weather is favorable, grill sales will spike a week or two ahead of the holiday as well as the actual holiday weekend. Given the highly seasonal nature of grill sales, demand planning and management are difficult within this product assortment.

Demand forecasts in the grill industry are often based on qualitative estimates by the merchandise buyers and store managers. Unfortunately, forecasts in this category are notoriously inaccurate. Industry averages show that the mean absolute percent error (MAPE) associated with individual stores on specific grill SKUs is more than 90 percent with a six-month planning horizon. Accuracy does improve somewhat at the corporate level with approximately 20 percent MAPE in the industry with a six-month lead time. Consistent with any type of forecast, accuracy in the grill industry increases in shorter time intervals. Megamart demand forecasts are similar to the industry averages reported previously.

The physical characteristics of gas grills present challenges for both Megamart distribution centers (DCs) and store locations. Gas grills are large, bulky, and difficult to handle. They ship partially assembled in large boxes that must be oriented in a specific manner and they are floor-stacked on import containers for shipment. From a material-handling perspective, the short-distance movement of grills requires either specialized equipment or additional labor. Because the grills are not palletized, utilizing traditional forklifts often damages the product with punctures. Therefore, distribution centers and stores must utilize nontraditional squeeze clamp forklifts or move the grills with a manual, two-person lift. The bulky nature of grills also presents a storage challenge in Megamart locations. Inside the distribution centers, grills do not efficiently fit in standard racking, so they are either bulk-stacked on limited floor space or immediately cross-docked to stores. In retail locations, shelf space is limited, so grills are stored in several labor-intensive ways. They can either be placed in "top stock" high above the retail selling floor (which requires multiple moves) or assembled and placed in front of the store (which requires additional manual labor). As these examples highlight, the movement and storage of grills within a Megamart location require additional labor, specialized equipment, and unique processes.

Stakeholder Concerns

A number of key stakeholders within Megamart have very valid concerns and conflicting goals when it comes to grills. Within store operations, managers want to have sufficient grill inventory available to meet customer needs and allow stores to exceed their sales budgets. As the primary interface with customers, stores expect to be nearly 100 percent in stock at all times on hot-selling items. Store managers not only want to be in stock, but they also want to have sufficient depth of inventory to fill their product displays, merchandise assembled grills, and demonstrate to customers that "*Megamart is THE destination for all grilling needs.*"

Early in the selling season, stores are often very vocal about trying to obtain some extra inventory a few weeks early so that they are ready when spring arrives, the weather breaks, and sales take off. Some stores even begin hoarding inventory so they have plenty if supply shortages occur. At the end of the selling season, store managers often want to return unsold inventory to the distribution centers or move products to other retail locations so that their individual store budgets do not have to incur margin losses for marking down products to sell in the off-season. The "hoarding" mentality going into the selling season quickly changes to a "dumping" mentality out of season. Mr. Manrodtner has already received multiple calls from the senior vice president of store operations regarding the necessary inventory levels for next year and how "he better not screw this up."

Distribution centers are another stakeholder group that can be dramatically impacted by grills. Megamart DCs are highly mechanized and efficiently move the majority of products via a state-of-the-art conveyor system. Unfortunately, due to their size and weight, grills are not conveyable. Movement of grills requires squeeze clamp forklifts and the DCs only have limited quantities of these specialized material-handling devices. Therefore, distributing grills takes much longer for DCs than a typical conveyable boxed product.

Compounding this handling problem is the seasonal nature of grills. DC managers simply do not have enough capacity to handle large quantities of grills in condensed time periods during the peak weeks of holiday sales. DC managers have repeatedly asked for grills to flow more evenly into retail locations rather than trying to ship just-in-time for peak sales weeks. Without a more even flow of grills, DCs will continue to experience costly bottlenecks during the critical spring and summer selling season. Such backlogs affect more than just the grill assortment. An entire DC can become backed up with grills due to the large cubic volume and the need to immediately unload and turn around import containers. When these types of delays occur, all products suffer service-level issues and the stores miss sales plans. The senior vice president of distribution has also contacted Mr. Manrodtner to convey his concerns about product flow plans for next year.

The merchandise planning and replenishment group views grills as the most difficult category to manage in all of Megamart. Normally, the group deals with domestically sourced items with short-order cycle times and relatively flat, predictable demand profiles. Maintaining in-stock service levels, sales plans, and inventory turnover objectives is relatively easy in other categories with these characteristics. However, grills are very different. They have long order cycle times and short, intense, highly seasonal demand profiles. The combination of long lead times and erratic demand patterns is especially difficult to manage because the order cycle time is nearly as long as the active selling season. Given this relationship, it is difficult to respond to changing consumer preferences and the group often relies on simple allocation heuristics.

Unfortunately, allocating products six months in advance of actual demand information usually leads to suboptimal results. Simultaneously maintaining store service levels and staying within corporate inventory turnover mandates is extremely difficult in the grill category. Typically, grills only achieve one of these two objectives. In the past, Megamart maintained high in-stock levels by purchasing too much inventory and pushing it all out to stores. Strong sales in season were then offset by dramatic markdowns of excess inventory, high inventory carrying costs, and excessive transshipment costs. Last year, Megamart attempted to maximize inventory turnover with a just-in-time approach to grill replenishment resulting in massive out-of-stock levels and gridlock in the distribution centers.

The Emotional Nature of Grills

Management of the grill supply chain is one of the most heavily scrutinized processes at Megamart. Various stakeholder groups are often pitted against each other and heated exchanges often occur between managers in various functional areas within the company. Finger pointing begins whenever a group begins to miss their annual objectives and performance bonuses. Store managers know that grills are one of the largest ticket items sold within a retail location.

If grills are not in stock at critical times during the year, a store will probably not make its overall sales plan. Therefore, stores can be highly critical of merchants who did not order enough grills or distribution centers that cannot ship to stores on time. Distribution center managers know that grills represent the largest cubic volume assortment within the entire company. If imported grill containers land all at once in a DC, throughput and service-level metrics will suffer. Therefore, distribution centers often criticize the way orders are placed by merchants and tell the stores to quit unreasonable, instantaneous shipment expectations.

Merchandise planning and replenishment managers need to ensure corporate financial objectives are met in terms of sales budgets, margin plans, and inventory turnover requirements. If they order too much, too little, too early, or too late, financial metrics will suffer. Merchants often complain about stores hoarding inventory and causing out-of-stocks or markdowns. They also claim distribution centers are extremely inflexible and cannot expect perfectly balanced inbound flow of highly seasonal categories. As these examples demonstrate, grills are a highly visible and emotional product assortment at Megamart. It seems like all eyes are now on Steve Manrodtner and his supply chain planning group to deliver a good plan.

Past Failures

Megamart has repeatedly attempted to "fix" the grill supply chain with a number of initiatives that have often failed to achieve one or more key goals. One year, all stores received identical direct shipments of import containers in order to bypass the Megamart distribution centers and avoid the annual spring gridlock. Unfortunately, not all stores had sufficient sales volume to handle full containers of a single grill SKU while other stores needed additional inventory. This initiative resulted in misplaced inventory, poor service levels, high markdown costs, and excessive transshipment costs. Some stores were completely out of stock and other stores had excess inventory. Service levels dropped below 80 percent, markdown costs exceeded 25 percent, and transportation costs increased 50 percent due to all the transshipment attempts to rebalance store inventory. Losses associated with this strategy exceeded $2 million.

Another year, the distribution centers decided to cross-dock all grill SKUs and push inbound inventory to the stores regardless of whether or not the stores needed additional inventory to support current sales. This initiative also resulted in inventory being pushed to the wrong locations so some stores had too much inventory and other stores did not have enough. Although distribution costs were reduced by 10 percent, markdown costs increased 20 percent and transportation costs increased 25 percent. This flow plan resulted in more than a $750,000 loss.

Last year, Megamart attempted to use a just-in-time approach and have import containers arrive at the distribution center right before peak season and then allocate inventory based on more up-to-date demand information. Unfortunately, these shipments slammed the distribution centers all at once and created costly bottlenecks that resulted in grills not being in the stores at the right times. Distribution costs increased 20 percent due to overtime rates, temporary employee increases, and the use of costly third-party logistics services. Service levels dropped in *all* store categories due to the congestion at the distribution center and many stores were out of stock on key items. Estimates show that stores experienced more than $20 million in lost sales because they were not replenished appropriately during key periods of time.

None of these attempts to "fix" the grill supply were acceptable to Megamart's senior management or customer base. The previous vice president of supply chain planning was fired after these failures. Mr. Manrodtner is feeling the pressure to succeed and he is counting on your group to develop a solution. There must be a better way to manage grill demand, flow the products, and ensure that cost/service objectives are met.

Relevant Details

To assist your team in its analysis on this project, the company has provided some relevant information about the grill category. First, detailed point-of-sales information is provided for the four grill SKUs under evaluation in the Excel data file (available from the Downloads tab on this book's website, www.ftpress.com/title/9780133448757). Item-level sales information is provided for each store in the pilot program by week for approximately a two-year period between the spring of 2009 and the spring of 2011. This rich source of data will permit your team to analyze sales rates; develop demand profiles and forecasts; perform A, B, C analysis of store types; and estimate purchase quantity needs.

In Table 1, a description of each SKU is provided along with information regarding individual product dimensions and how many grill SKUs will fit on a standard 40-foot container. The number of grills per container is based on a single SKU container and does not account for containers that may have more than one SKU loaded. In the past, the supplier would charge a 20 percent premium for mixed containers because it is difficult to load numerous configurations of mixed SKU orders. However, the supplier is now willing to charge the normal product price for a standardized mixed SKU container if Megamart provides this *single* standardized configuration well in advance, for planning purposes. This change gives Megamart some much needed flexibility to potentially ship higher-volume stores a direct shipment prior to peak sales periods and relieve distribution center congestion. The key will be to identify a standard configuration that a majority of stores could use to support sales.

Your Task

Apply basic demand management principles to the Megamart grill assortment and develop a detailed purchase plan for the company that appropriately flows grills throughout the supply chain and takes into consideration all key stakeholder concerns.

As you analyze data and develop your plan, you may want to consider the following questions:

1. What is the demand profile for grills?
2. How many units of each grill need to be purchased to support sales?
3. What is the demand profile for each SKU?
4. How many 40-foot container shipments will Megamart need?
5. How many 40-foot containers will land at the DC next year?
6. How many 40-foot container shipments should bypass the DC and go directly to individual stores? How much cubic volume would this remove from the DC?
7. How many 40-foot containers will land at the DC for each week of the year?
8. Is there a standardized, mixed SKU configuration that makes sense for direct store shipments? If so, what is the SKU mix (i.e., 10 units of X, 25 units of Y, 20 units of Z)?
9. Which stores will receive direct shipments and when will they receive them?

10. Which stores are the A, B, C volume stores? Will you treat them the same or differently?
11. Should grills be "pushed" or "pulled" to final store locations? Is there a hybrid approach?

The final deliverables for this project include an executive summary of key findings, a presentation to senior management explaining your recommendations and justifying why they are appropriate, and a detailed purchase plan that can be easily implemented by the replenishment group. Mr. Manrodtner is counting on your expert analysis and advice.

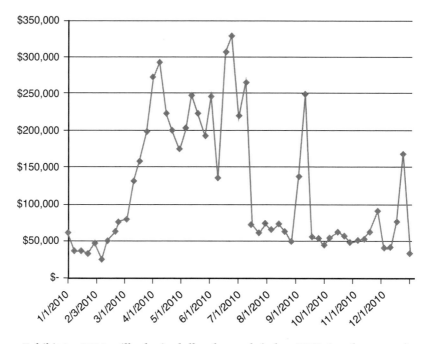

Exhibit 1   2010 grill sales in dollars by week (select SKUs in select stores)

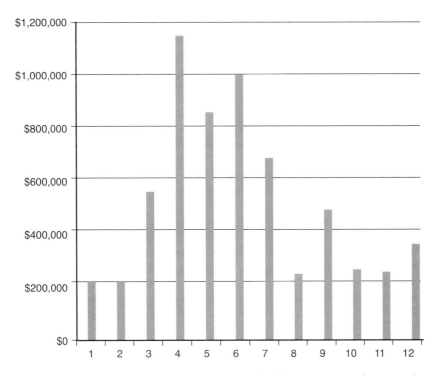

Exhibit 2   2010 grill sales in dollars by month (select SKUs in select stores)

Table 1   SKU Information

SKU Number	Description	Retail Selling Price	Length (Feet)	Width (Feet)	Height (Feet)	Number of Grills per 40-Foot Container
1099	2-Burner Gas Grill	$99	3.5	2.5	3	66
1199	4-Burner Gas Grill	$199	4	3	3	36
1399	3-Burner Premium Gas Grill	$399	5	3.5	3	32
1499	4-Burner Premium Gas Grill	$499	5	3.5	3.5	32

4

SUPPLY UNCERTAINTY, DEMAND PLANNING, AND LOGISTICS MANAGEMENT AT GOODWILL INDUSTRIES OF OKLAHOMA

Chad W. Autry, Ph.D.

Introduction

When Heather Robertson and Chris Davis arrived for their weekly lunch at the Spaghetti Depot in the heart of Oklahoma City's trendy Bricktown district, Heather's mood was spirited, as always. Chris, on the other hand, seemed somewhat weary.

Since Heather had taken over as CEO of Oklahoma Goodwill Industries (OGI) in March of 2008, change had become the status quo. Within months of her hire, two of OGI's 12 Goodwill retail shops had been closed, and were replaced by three smaller, leaner collection-only facilities—"attended donation centers" where a donor can make drop-offs but no merchandise is sold. Inventory counts in the remaining 10 stores are more accurate than before, and merchandise turnover is robust. So are sales for the territory, which had eclipsed $7 million for the first time in recent history.

As they began to eat, Heather was quiet but smiling—her normal state—while Chris played moodily with his tortellini appetizer. After assuming the position of vice president of operations, he had gradually become Heather's right-hand man and their weekly lunches turned into cathartic experiences—a time when they could decompress and look at OGI's business issues from a distance. It was there that he felt most free to openly share his ideas and concerns with his relatively new mentor.

This week, as usual, he was armed with some of both. Under her direction, he had recently completed a new five-year, forward-looking plan, designed to wrestle the nonprofit's supply-and-demand planning and to get logistics management processes under control. Although he was eager to share the details, there were clearly a number of concerns on the horizon.

As a nonprofit organization that relies almost exclusively on donations to stock its retail locations, OGI's inventory is virtually free from a product cost standpoint. However, forecasting inventory volumes and sales across product categories is very difficult; thus, marketing decisions are very tricky to make, and operations costs are deceptively high as a percentage of goods sold. In addition, even though the complexity and costliness of the OGI supply chain had been somewhat reduced by replacing the two underperforming retail stores with three attended donation centers, the action also reduced OGI's market presence. OGI now operates only 15 percent of the total nonprofit consumer goods retail floor space in the Oklahoma City metro area, down from nearly 20 percent in 2007, making Chris begin to worry that customers' reduced store access—along with the associated reductions in donor and customer brand awareness—could become problematic. Donations per retail square foot are up since the Shawnee and Britton Road locations were closed, but total donation volume across the territory is relatively static.

Furthermore—and a hot button for Chris—OGI has invested in significant capital equipment in recent years, the fixed costs of which were now being spread among only 10 retail locations. Trucks were a major capital expenditure, as were retail racks, sorting bins, and the 160,000-square-foot distribution center (DC), which also housed the company headquarters and the Walker retail location. A significant amount of this equipment was underutilized and/or lying dormant and Chris had quietly begun to worry whether the recent closings hadn't led to more costs than expected. Trucks in particular had been running at only 73 percent capacity, down from 88 percent the previous year. Certainly the economic downturn of late 2008 had negatively impacted donations to some extent, but surely there was a better way than this for OGI to leverage its logistics assets.

For Chris, this month was an opportunity to both improve OGI operations and impress his new boss. Thus far, Heather had been great to work for, and he had learned a lot from her experience of having turned around a mediocre territory in her home country of Canada in the previous five years. Nevertheless, he retained a haunting feeling that the challenges that lay ahead could be greater than either of them had yet faced.

He took a gulp of his honey-sweetened iced tea and began to share his concerns. She nodded knowingly, empathetic to his concerns, and sat back thoughtfully as he laid out a bullet-point sketch of his proposal. As he concluded, she agreed that there was much work remaining to be done with respect to both demand planning and logistics operations. However, she was confident that they could come up with a solid game plan—one that would not only include new methodologies for addressing OGI's supply-and-demand uncertainties, and allow for more effective use of logistics-related capital assets, but could also lead to the opening of an additional new retail location that would facilitate the access and name recognition that Chris was convinced OGI was in danger of losing. However, time was of the essence; their report to Goodwill International on the status of the territory was due in less than 60 days, and the clock was ticking. She asked Chris to flesh out his proposals in much greater detail, and report back to her in one week—which left plenty of time for firm commitments to be made and if adopted, be included in the territory status report.

Oklahoma Goodwill Industries (OGI)

For more than 100 years, Goodwill Industries has been a leader in using the power of work to enrich individuals' lives. In 1902, a Methodist minister, the Reverend Edgar J. Helms, had just completed his seminary training and arrived in his new parish on the south side of Boston. This section of the city was populated with European immigrants who, unfamiliar with the English language and running out of money, were losing hope of the American dream. Helms began collecting household goods and clothing from the wealthier areas of Boston. Because the immigrants were too proud to accept donations, he hired and trained the poor to repair and mend the items. These items were then either sold or given to the people who repaired them. This system was a success and the Goodwill philosophy of a "hand up, not a hand out" was born.

By the mid-1930s, E. K. Gaylord, founder of the Daily Oklahoman newspaper in Oklahoma City, had heard about the Goodwill philosophy and his interest was piqued. He sent his highly acclaimed journalist, Edith Johnson, on a fact-finding mission to visit the St. Louis and Ohio Goodwill operations. Edith returned to Oklahoma convinced that a Goodwill operation should be incorporated in the "Capital City."

The formal incorporation for Oklahoma Goodwill Industries occurred on September 21, 1936, and operations began soon thereafter in a two-story building at 11th and Robinson Street, near the growing downtown area. By 1960, the Oklahoma Goodwill operations had moved its headquarters to a location inside the former bus terminal at 410 W. 3rd Street and additional real estate lots were purchased around the city in anticipation of future growth.

Fast forwarding to 2009, Oklahoma Goodwill Industries has 10 retail locations, totaling more than $7.7 million in 2008 retail sales. The OGI territory covers the surrounding towns of Stillwater, Shawnee, Yukon, Midwest City, Norman, Edmond, and Moore, as well as three locations within Oklahoma City proper. Most important, OGI is currently

Oklahoma's largest nonprofit provider of employment and training services for people with disabilities and other disadvantages. Thus, OGI promotes itself as "not charity, but a chance."

OGI's business model relies on donations of clothing, household goods, furniture, electronics, and other items as sources of inventory to stock their retail stores. There are three sources of revenue: retail sales, secondary sales, and contract services that contribute to the organization's bottom line. Donations that are dropped off at retail locations are typically sold in the same location. Prior to being merchandised, the donations, which often take the form of large bags, boxes, containers of household goods, or even a donor's pickup truck (filled with discarded garage and lawn equipment), are opened and sorted according to the receiving store's inventory needs, by product type, and by usable condition—all judgments are made by the retail store's general manager and his/her staff and based on experience.

Based on these criteria, items deemed unusable or unwanted are consolidated and sent to the Walker location distribution center for further evaluation. At the Walker DC, usability judgments are again made, with sellable items redistributed to stores (other than the sender) that have expressed a need for the product. Items deemed unsellable at the DC level are either packaged as bulk material to be sold into the secondary marketplace (i.e., raw fabrics or electronic components markets), or in the case of extreme malfunction/lack of value, are either recycled or disposed of.

Sales into the secondary markets comprise approximately one sixth of OGI's overall revenue. In addition to the sales streams, a newly piloted segment of OGI's business consists of the contracting of basic services, such as maintenance, custodial, and landscaping, to state and municipal agencies. OGI bids on any government contracts that become available in these and similar areas and staffs the contracts they have won with crews composed of persons in need from the local community. Thus, OGI serves as a de facto employment agency for a portion of its own customer base. Many of the OGI contract employees are physically or mentally handicapped; they may also be individuals who have been out of work for an extended period of time and are in need of financial assistance while searching for a permanent position. These government bids provide Goodwill the opportunity to help place individuals into more stable and permanent employment situations rather than leaving them to rely solely on state programs for food and shelter.

In 2008 based on this business model, OGI sketched out an aggressive five-year plan, designed to increase donations and revenue. The plan was based solely on improved logistics and supply chain operations. To highlight the plan, 2008 revenue was approximately $7.7 million, and expectations for 2013 are $13.9 million; thus, an increase in revenue of approximately $6.2 million is put forth as the primary OGI objective over the next five years. To accomplish such a goal, it appears necessary to increase donations/square foot by approximately 40 percent per store, on average.

However, as Chris began to consider these goals, it was becoming increasingly unclear as to precisely how he should go about achieving them. Three different bundles of questions keep coursing through his mind:

- How can OGI gain a greater understanding of its inventory's locations and handling throughout the organization? That is, how can it get needed items to the right store location at the right time and in the right sellable condition, when it has very little idea as to what products are coming into the organization through donations in the first place?
- Are the trucks owned by OGI being used to their fullest capability? Should some trucks be sold, additional trucks purchased, or on that note, should transportation be outsourced altogether?
- Are the attended donation centers revenue generating and cost effective, or are they more trouble than they are worth? Are they in the right locations? And/or, should OGI consider adding an additional retail store in a more appropriate location in order to better serve the community?

To address OGI's concerns, Chris immersed himself for nearly two months in the task of learning more about the areas of logistics and supply chain management through extended education courses, networking with area operations managers from other industries, and where possible, benchmarking other nonprofit organizations. Following these activities, as well as attending three meetings sponsored by his local CSCMP roundtable, he was able to narrow his research to two areas of focus: supply uncertainty/demand planning and logistics management (with the latter specifically focused on transportation and facility location aspects).

It was at this point that, anticipating an upcoming lunch with Heather, he turned his attention to their next meeting, and to the annual report to Goodwill International soon to follow.

Supply Uncertainty and Demand Planning Issues at GIO

Generally, a for-profit company will have a relatively firm grasp on its inventory management, distribution, transportation, and marketing processes because it controls its procurement and ordering functions that feed inventory into the system. This arrangement enables the for-profit business to control its internal and outbound supply chain processes because within some reasonable margin of error, it knows what products or commodities are inbound and when they will be arriving. However, not-for-profit retailers such as Goodwill are unique in this regard because they don't order products based on some periodic or reorder point–based heuristic; rather, they are dependent on donations received at random for stocking their stores with inventory. Therefore, inventory management, transportation management, marketing, and distribution are more difficult to plan for in the not-for-profit scenario. Though basic sales forecasting is possible, inventory forecasts are subject to wild variance swings from period to period and product type to product type. As one Goodwill executive once explained to Chris, "…we generally have some idea as to *how much* we are going to sell; we just have no idea exactly *what* we are going to sell."

To some extent, the mysteries surrounding demand and supply at OGI are related to the way it accounts for and accepts donations received from donors. At OGI, a donation is defined internally as the receipt of a bundle of unwanted goods from a donor, regardless of content, mix, or size, resulting from a single drop-off or pickup. In other words, one donor could bring a truckload of used books, lamps, and furniture to the Walker street distribution center, while another could request home pickup of a small sack of slightly worn clothing, yet both would be credited with having made a single donation. This process was born from the necessity to make the donating process as easy and quick as possible from the perspective of the donor; donors don't want to wait around while each of their items is sorted, examined, categorized, and catalogued into an inventory management system. In fact, many of the donors who want to gain a tax credit for their charitable activity still prefer that the specific content of their Goodwill donation remain somewhat private or anonymous.

Chris and Heather have often discussed whether different definitions and accounting for donors would be desirable, but neither has been able to identify a solution that justifies the additional inconvenience to the donor (the "lifeblood" of the OGI organization) that alternative systems almost necessarily create.

However, though this system is effective at increasing donation frequencies, it renders a number of complexities for the retail operation. Because items are received in mixed containers of unpredictable volumes received at seemingly random time across a universe of stores within the territory, no viable way of creating store-to-store (and in some cases, intrastore) inventory visibility has yet been discovered. Furthermore, due to the very low average sales price per item, the application of both traditional and emerging inventory technologies (e.g., bar coding and/or RFID) seems cost prohibitive to Chris. For example, the average unit of clothing inventory sells for approximately $2.52, with around $1.75 of the sales price tied up in labor, utility, and retail leasing costs. The proportional increase in costs associated with implementation of inventory technology would therefore swallow up much of the remaining margin, assuming 2008 per unit technology prices. Chris is currently struggling with the decision to invest in supply chain technology as a methodology for increasing inventory visibility, but given that he knows little about the technologies or their implementation, he seems to be leaning against it, and is searching for a less costly, more immediately effective solution.

In addition, the volatility in the OGI donation supply market is accompanied by time-based complications in the demand market. For instance, donated items with a seasonal supply component are often received immediately following the season in which they are most useful (having been used then by their donors). Thus, customers shopping OGI retailers will tend to demand products that are useful in the near future, while the store's inventory is full of supply most useful in the recent past. This means, for instance, that though a fair number of customers will shop an OGI retailer for a winter coat during the month of October, the racks are fairly likely at that time to hold only a few winter coats, while at the same time holding an excessive number of swimsuits, t-shirts, and other summer wear. Goodwill has yet to develop a demand management system capable of eliciting certain types of donations during and immediately prior to the season when their sale will be most likely and profitable. Marketing efforts directed toward supply and demand balancing in the event of lagging sales for a particular item have traditionally been ignored by the company given the inherent randomness of the acquired inventory.

Both Chris and Heather believe that this sort of supply-demand misalignment is simply a necessary but unfortunate obstacle that must be dealt with by nonprofits, though they each express willingness to be convinced otherwise. However, other Goodwill corporations in distant parts of the United States and Canada have experimented with demand solicitation in advance of seasons with some modest success.

Logistics Management Issues at GIO

In addition to supply uncertainty and demand planning issues, Chris is also convinced that a key element of OGI's future success is the more effective utilization of the capital assets OGI has devoted to logistics. In this regard, two elements are of primary concern to both him and Heather: the use of three OGI-owned trucks and the long-term leases held for the (currently 13) retail shops and attended donation centers. A secondary concern is the $37,500 worth of retail fixtures, aluminum racks, storage bins, cash desks, and electronic surveillance equipment collected from the two now-closed retail locations that currently sit dormant in the Walker DC.

Currently, the three trucks are running at 73 percent aggregate capacity. If OGI can implement a plan to increase truck capacity utilization, efficiencies will be gained and OGI will be in a better position to achieve the long-term goals set out in the plan. Each of the three trucks in the OGI fleet holds an average of 50 donations when loaded at full capacity. Two of the three trucks were purchased in 2007 as replacements for previous trucks already sold at salvage; each cost $72,000 and has a total eight-year depreciation basis. The other truck, purchased in 2005 for $64,000, is in the fourth year of a nine-year depreciation basis. One of the trucks (A) is devoted solely to home pickups of donations, which are scheduled roughly five to seven business days in advance. The remaining two trucks (B and C) run daily routes between the Walker DC, the retail stores, and the attended collection centers. If home pickups are scheduled and are located roughly along these routes, then these are incorporated into the B and C trucks' daily schedules; all other home pickups are handled by truck A. Based on the data, Chris's first inclination is to sell one of the trucks and spread the current capacity among the remaining two. However, if donation demand increases as expected in the five-year plan, this could possibly leave OGI with insufficient capacity to transport its inventory by 2011.

Additionally, comments made by Heather at the most recent month's managers' meeting indicate that she views the closure of the Shawnee and Britton locations as steps toward optimizing the OGI retail network. To this end, she has asked Chris to collect the necessary data to determine whether OGI would benefit by adding an additional retail store and/or attended collection center(s). Though he is not sure that such a maneuver would be supported in the short term by significant added sales, he has always believed that OGI needed the additional retail presence in order to defend its territory from a brand-awareness perspective. Therefore, he is interested in finding two to three potential locations where future stores could effectively be established (assuming, of course, that space is available). Furthermore, both Chris and Heather are wondering whether the attended collection centers are adding value to the OGI supply chain. Unfortunately, neither of them has the requisite experience in the area of supply network optimization to allow them to make an adequate assessment.

Concerns for the Immediate Future

Chris's proposals are due to Heather in one week, and given his limited experience in the areas of logistics and supply chain management, he needs any assistance that can be acquired. He has sought out your assistance as a logistics/supply chain consultant in addressing the issues currently facing OGI. Any information and/or wisdom that can be provided will be viewed as a helpful, positive step toward cementing the relationship between yourself and OGI, and could lead to additional future business between you and the organization.

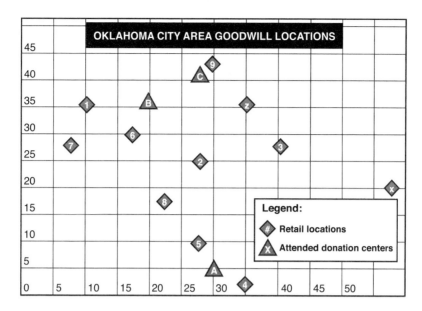

Notes: Retail location #10 (Stillwater) is not shown on map. Locations X and Z represent the recently closed Shawnee and Britton Road retail locations, respectively. Coordinates are scaled in miles.

Location	Type	X-Coordinate	Y-Coordinate	Location	Type	X-Coordinate	Y-Coordinate
1—Council	Retail	10	35	8—Penn	Retail	23	18
2—Walker	Retail	28	25	9—Edmond	Retail	30	42
3—Reno	Retail	41	27	10—Stillwater	Retail	53	80
4—Norman	Retail	35		A—North Norman	ADC	30	5
5—Moore	Retail	27	10	B—May	ADC	20	35
6—MacArthur	Retail	18	29	C—Kelly	ADC	28	39
7—Yukon	Retail	7	27				

Exhibit 1 Goodwill Industries of Oklahoma: retail store and attended donation center locations

Case 4 Supply Uncertainty, Demand Planning, and Logistics Management

Table 1 Monthly Sales, January 2008–December 2008, by Location, in Dollars

Location	J08	F08	M08	A08	M08	J08	J08	A08	S08	O08	N08	D08	2008	Avg.
Council	42,490	33,489	40,310	36,418	39,078	38,858	44,520	37,807	36,629	39,927	43,225	46,523	479,274	39,940
Walker	20,682	20,997	20,413	19,001	18,351	18,115	19,664	21,037	18,930	22,228	25,526	28,824	253,768	21,147
Reno	103,645	97,369	107,405	95,998	83,887	84,422	91,732	96,673	99,138	102,437	105,735	109,033	1,177,474	98,123
Norman	38,448	37,269	39,190	35,569	39,178	39,880	41,857	41,824	40,080	43,378	46,676	49,974	493,323	41,110
Moore	39,213	35,517	38,313	34,039	38,691	30,469	42,493	44,047	38,637	41,935	45,233	48,531	477,118	39,760
MacArthur	53,394	47,962	54,536	48,385	50,398	52,207	60,858	62,177	60,589	63,857	67,175	70,483	692,021	57,668
Yukon	24,289	23,891	27,458	24,474	25,003	24,832	28,608	30,324	29,316	32,615	35,644	39,211	345,665	28,805
Penn	82,158	82,909	94,945	90,069	99,276	91,472	95,639	100,536	91,459	94,757	98,056	101,354	1,122,630	93,553
Edmond	45,963	45,561	47,055	45,723	46,545	57,932	64,120	58,621	62,441	65,739	69,037	72,335	681,072	56,756
Stillwater	24,658	25,690	29,937	26,269	28,641	25,904	26,446	29,906	27,606	30,905	34,203	37,501	347,666	28,972
Shawnee	38,776	38,425	44,117	39,552	38,177	38,920	36,749	33,115					307,831	38,479
Britton Rd	42,282	38,197	43,375	38,715	39,046	39,799	42,066	37,764					321,244	40,156
TOTALS	555,998	527,276	587,054	534,212	546,271	542,810	594,752	593,831	504,825	537,778	570,510	603,769	7,752,221	553,729

Table 2 Monthly Donations, January 2008–December 2008, by Location, in Numbers of Donations

Location	Ft^2	J08	F08	M08	A08	M08	J08	J08	A08	S08	O08	N08	D08	2008
Council	19,600	729	311	467	531	677	570	710	725	1,106	834	876	926	10,346
Walker	8,800	399	396	140	144	174	424	403	375	364	368	356	350	4,592
Reno	25,775	933	705	1,083	1,101	1,610	2,115	1,822	1,539	1,568	1,630	1,566	1,576	20,450
Norman	18,995	1,045	824	979	911	1,326	1,158	1,292	1,046	808	1,036	950	918	13,988
Moore	17,770	770	588	735	791	970	834	1,006	899	942	936	913	918	12,138
MacArthur	20,220	1,440	1,313	1,760	1,795	2,380	2,292	1,659	1,125	1,152	1,308	1,191	1,204	21,046
Yukon	15,900	728	471	828	787	765	794	796	965	862	866	884	856	11,333
Penn	23,000	912	714	993	1,040	1,232	1,227	1,201	1,262	1,010	1,145	1,126	1,081	15,335
Edmond	20,000	892	627	810	1,139	1,035	1,143	1,284	1,410	1,217	1,291	1,293	1,004	14,793
Stillwater	12,000	611	499	542	685	921	763	791	795	593	714	688	652	9,514
Shawnee	10,500	201	141	197	269	366	329	348	255					2,106
Britton Rd	20,400	706	646	600	660	759	788	791	608					5,558
NN										256	280	251	191	1,783
May										408	392	390	333	2,437
Kelly										272	380	399	425	2,504
Home Pickup		614	474	599	646	801	816	794	722	692	606	544	567	9,078
TOTALS	21,2960	9,980	7,709	9,733	10,499	13,016	13,253	12,897	11,726	11,250	11,786	11,427	11,001	157,001

Table 3 Monthly Secondary Market Sales and Recycling/Refuse Counts, January 2008–December 2008 (Walker DC)

Location	J08	F08	M08	A08	M08	J08	J08	A08	S08	O08	N08	D08	2008	Avg.
Secondary	7,671	7,274	8,225	7,436	7,625	7,550	8,312	8,288	8,410	8,971	9,506	10,005	99,402	8,863
Recycling/refuse	157	203	144	157	272	588	904	431	489	555	691	703	5,294	441.1

Note: Secondary sales are in dollars; r/r counts are in donations.

Table 4 Truck Routes and Monthly Donations Handled, January 2008–December 2008, by Truck

Truck	J08	F08	M08	A08	M08	J08	J08	A08	S08	O08	N08	D08	2008	Avg.
A	598	445	580	613	755	801	734	692	638	583	507	540	7,486	623.8
B	806	605	797	890	1,115	1,134	1,107	1,013	998	925	1,002	1,056	11,448	954.0
C	771	630	744	785	967	954	970	851	816	961	849	710	10,008	834.0
TOTALS	2,175	1,680	2,121	2,288	2,837	2,889	2,811	2,556	2,452	2,469	2,358	2,306	21,456	1,788.0

Notes: Route A = home pickups; Route B = 2-8-5-A-4-3-2 plus homes if needed; Route C = 2-6-7-1-B-C-9-2 plus homes if needed

Appendix

Customer donation centroids (areas/neighborhoods) and expected donations, 2009–2013:

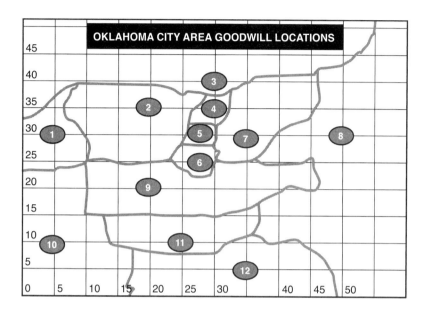

Neighborhood/ Area	Don. Centroid	2009	2010	2011	2012
1. Yukon/Mustang	(5,30)	11,700	12,800	13,500	14,100
2. Warr Acres	(20,35)	8,000	8,400	9,250	9,575
3. Edmond/Deer Creek	(30,40)	15,750	16,900	17,700	18,250
4. Village	(30,35)	7,500	8,300	8,700	9,125
5. Nichols Hills	(27,30)	20,500	25,900	30,750	36,600
6. DT/Heritage	(28,25)	5,500	5,800	6,350	6,575
7. State Capitol	(35,30)	3,550	4,500	4,900	5,400
8. East Burbs	(50,30)	22,570	22,800	22,900	23,400
9. Southside	(20,20)	16,200	19,300	21,700	25,575
10. Wheatland	(5,10)	1,200	1,350	1,470	1,550
11. Moore	(25,10)	13,250	18,300	21,900	24,700
12. Norman	(35,5)	14,500	15,550	17,000	18,500

Part 2

SUPPLY CHAIN NETWORK DESIGN AND ANALYSIS

This section contains four cases: (1) "Bertelsmann China—Parts A and B: Supply Chains for Books," (2) "Carnival Corporation Food Supply Chain," (3) "DSM Manufacturing: When Network Analysis Meets Business Reality," and (4) "Kiwi Medical Devices, Ltd.: Is 'Right Shoring' the Right Response?" Although the focus of the four cases is best described as supply chain network design and analysis, each case examines a particular aspect of network design and analysis, which particularly challenges managers.

The "Bertelsmann China—Parts A and B: Supply Chains for Books" case, which takes place in China, focuses on one of the world's leading media companies and illustrates its constant need to redesign its supply chain in an extremely fast-growing market, which renders formerly developed layouts as no longer applicable. It is a two-part case wherein the individual parts stand alone, yet can be considered together as they address different aspects of the network design topic. Bertelsmann China Part A demonstrates the importance of push/pull boundaries in supply chain network design and emphasizes the high-impact strategic decision making of such a challenge. Cost and performance drivers are prominent components of the case, as well as challenges faced by differences in culture. Bertelsmann China Part B demonstrates the importance of costs (as well as lead time and responsiveness) in a low-margin business sector, and thus linear programming methods are applied to optimize the network.

The "Carnival Corporation Food Supply Chain" case focuses on the need to identify strategies to significantly reduce food supply chain costs, while maintaining quality and customer service. The case demonstrates unusual challenges and constraints that are unique to the cruise line industry, in particular the inability to replenish the supply chain for a one-week time period. The case requires students to analyze and interpret benchmark data when formulating recommendations; it also requires the evaluation of the merits of various supply chain design improvement tactics.

The "DSM Manufacturing: When Network Analysis Meets Business Reality" case focuses on a network optimization project and the need for a consulting team to generate a recommendation for a contract manufacturing location for a product. The case demonstrates the importance of understanding additional trade-off choices within a supply chain that extend beyond a straightforward cost analysis. It also illustrates the challenges that companies face when they globalize operations and expand their customer base. Also, it points out that the scope of a project can lead to potentially incorrect answers. The case differentiates itself because it is well grounded in reality: It suggests that the creation of constraints is often a result of company politics rather than a pure desire to find the best solution.

The "Kiwi Medical Devices, Ltd.: Is 'Right Shoring' the Right Response?" case focuses on the competitive dynamics in today's global marketplace and the strategy of "right sourcing." The case requires identifying the various factors (beyond labor costs) that drive offshoring, consideration of country and operating mode trade-offs, and the diversity of decisions that must be made to ensure that an evolving global network design is capable of meeting a company's competitive needs. Weighted-factor analysis is illustrated as a tool to enhance sound decision making when a selection must be made among competing options.

5

BERTELSMANN CHINA— PARTS A AND B: SUPPLY CHAINS FOR BOOKS

Professor Stephan M. Wagner, WHU
Viviane Heldt, WHU
Katrin Lentschig, WHU
Jennifer Mayer, WHU

Bertelsmann China—Part A

"Ding, Dang, Dong." In a Bertelsmann bookstore in the city of Harbin, the doorbell interrupted shop manager Foguang Chou's typical Monday morning thoughts.

The Bertelsmann Book Club shop manager regained his poise quickly and found himself confronted by the same longtime customer who had asked him for the newest release of *Harry Potter* the previous week. Using his moist hands to wipe the sweat from his forehead, the experienced manager mumbled excuses and promised that the book would be available the following week. The customer was a loyal member of the Bertelsmann Book Club, but Foguang Chou could see that she was becoming increasingly impatient.

However, he knew how she felt. After all, who wanted to wait for a best seller that had already been advertised in the Book Club's catalog due to its unavailability in stores? If only this woman were the only customer waiting for the newest books to arrive. During all of June and July 2006, Foguang Chou had been busy finding excuses to give all of his customers about why the shipments of new titles kept arriving late.

Headquarters needed to ensure on-time delivery of the books that were advertised in Bertelsmann's monthly catalog. Foguang Chou would have to discuss this issue next week at the shop managers' conference so that the problems could be solved on a national level.

History of Bertelsmann

Bertelsmann AG was founded in Germany in 1835. The founder was the printer Carl Bertelsmann who first started a publishing business and then opened a book-printing plant in Gütersloh. By 2006, the company had grown from a pure-play publishing and printing firm into an international media conglomerate. In accordance with the slogan "strive to make media happen," Bertelsmann developed leading companies in many segments of the fast-moving media market. Mutual trust and responsibility between the employees and the company became enshrined in the "Bertelsmann Essentials." These principles of partnership, entrepreneurship, creativity, and citizenship had to be incorporated by each of the firm's employees.

The diverse business areas of Bertelsmann AG incorporate Europe's largest TV broadcaster (RTL, Luxembourg) in addition to the world's largest book-publishing group (Random House, New York). Bertelsmann AG has also established Europe's biggest magazine publisher (Gruner + Jahr, Hamburg), a media services division (Arvato, Gütersloh), media distribution (Direct Group, Gütersloh), and music publishing (BMG Music Publishing, New York). The Direct Group division, which is at the center of the case, is known in particular for its Bertelsmann Book Club. In 2005, Direct Group contributed about 13 percent to total revenues (Exhibit 1). Bertelsmann's operations spread throughout 63 countries and have approximately 88,500 employees. Gunther Thielen is the chairman and CEO of the firm. The Mohn family owns 23 percent of the shares and also controls the Bertelsmann Foundation, which holds the remaining 77 percent. Clearly, the holding family has a decisive influence on the corporation's values.

Bertelsmann Begins to Do Business in China

Bertelsmann attempted to join the Chinese media industry in 1993 when the former German chancellor, Dr. Helmut Kohl, made his historic visit to China. Bertelsmann, as a part of the German delegation, began negotiations with Shanghai authorities to discuss possible opportunities for cooperation. Their first business opened in China after completing negotiations in 1994. As doing business in China required licenses, a joint venture with the China Science & Technology Book Company was established in order to allow Bertelsmann to found the Shanghai Bertelsmann Culture Industry Company in Shanghai. In 1997, the company took another major step, creating the Bertelsmann Book Club (BBC) as the first joint venture of its kind in Shanghai. Due to the emerging middle class and increase of private wealth, in 2006, the Bertelsmann Book Club reached more than 1.5 million members with more than 7 million books, CDs, VCDs, DVDs, and computer games sold annually. By now, Bertelsmann operated in several business fields and had multiple subsidiaries in China, including Direct Group's Bertelsmann Book Club and 21st Century Book Chain Company.

Under the terms of its joint venture with the China Science & Technology Book Company, the Bertelsmann Book Club was only licensed to conduct business in Shanghai. Once Bertelsmann decided to open a club center business outside of Shanghai, aside from a mailing business, another joint venture was necessary. In June 2003, an official joint venture with Beijing-based 21st Century Book Chain Co., Ltd., was created to serve Chinese customers through Bertelsmann club centers outside of Shanghai. Thus in 2006, Direct Group China consisted of the two business units: 21st Century and BBC. BBC was responsible for the catalog business and operated 9 shops within Shanghai. 21st Century ran 28 shops in 16 Chinese cities.

Chinese Culture and Impact on Western Businesses

Doing business in China brought many challenges for Western firms. There were differences in corporate and country-specific culture, especially for the German-based Bertelsmann AG. In general, Chinese people relied on etiquette and the preservation of harmony, even if this meant that conversations were less than fully truthful. Confrontations were avoided to make sure that people would not lose face.

Power distance (as referenced by Hofstede) in the Asian culture was high, meaning that hierarchy was considered normal. Nevertheless, employees were not always acting in full loyalty toward their employers. This could partly be countervailed by providing long-term goals, which assured security and giving fixed vacations, partly in the form of departmental outings. Collectivity and social networks were extremely important, as were the individual business networks "guanxi" and "danwei," the groups whose members developed a corporate feeling. If there was a strong group feeling, colleagues would not blame each other for any mistakes. Instead, Chinese employees either blamed the government, which they could not control and did not trust, or they relied on superstition to relieve themselves of some responsibility.

Industry Overview of the Chinese Book Market

The main competition for Bertelsmann China came from low price discounts and from Internet companies that were more successful in gaining market share. Due to the increasing popularity of the Internet, Internet retail firms such as Amazon.com put intense pressure on the business performance of more traditional retail businesses. In China, the communist regime and its tradition of many state-owned local firms created additional challenges for international corporations such as Bertelsmann. Whereas the

Chinese competition could draw on local knowledge and close relationships to state authorities and formerly state-owned publishers, Bertelsmann could not.

Another major competitor of Bertelsmann China emerged in 2004 when the former editorial director at Bertelsmann Direct Group China took Bertelsmann's business idea and set up his own book retailing company, called "99 Read." With a better knowledge of the country and more personal relationships, 99 Read was able to build strong publisher relations that allowed them to source at more favorable terms than Bertelsmann could. Bertelsmann responded to the pursued discount strategy of 99 Read by selling exclusive titles after signing contracts with publishers.

Direct Group China's Supply Chain

The most important suppliers for Direct Group China were its publishers. They presented new books and authors to Direct Group China and were responsible for "manufacturing" the book. This "manufacturing" included all steps from writing to editing, printing, binding, and finally delivering the book to Bertelsmann's warehouses.

BBC and 21st Century were joint ventures that allowed Direct Group China to run shops all over China. BBC's business made up 75 percent of the combined BBC and 21st Century revenues. Looking at the shops from both business units, they made up 38 percent of combined BBC and 21st Century revenues and were growing faster than the mail business. For legal reasons, the two business units ran separate warehouses, both located in Shanghai. The entire catalog business and the nine BBC shops within Shanghai were supplied by the BBC warehouse, whereas the 21st Century warehouse was responsible for supplying the 28 stores in 16 cities outside of Shanghai.

The concept behind the club business demanded that every member be sent a catalog on a regular basis from which he or she could order via phone, fax, mail, or the Internet. Orders were usually processed within 24 hours and shipped to the customer using Chinese postal services. The lead time from the warehouse to the customer's home depended on the distance and could vary from one day to one week.

Supplying Direct Group China's shops worked in a similar way. First, the warehouse manager notified shops about available items. Two days later, the shops placed their orders. Shanghai shops placed orders twice a week and could be supplied within a business day due to their close proximity to the warehouse. The supply chain to China's other 28 shops was more difficult. Shops ordered only once a week, and, depending on their distance from the warehouse, the complete outbound logistics cycle from receiving the order notice of the warehouse manager to the arrival of books took six to ten days. This included one day for the third-party logistics provider (3PL) to prepare the order and three to seven days to ship it.

Catalog Development Process

The main catalog contained 48 pages and was issued at the beginning of every odd-numbered month. For instance, if the first catalog of the year was sent in the beginning of January, the next main catalog would offer a new mix of books and gifts in early March, and every two months thereafter (Exhibit 2).

To create the catalog, a complex workflow linking all responsible parties took place at BBC and 21st Century (Exhibit 3). The BBC catalog development process cycle started 65 calendar days before catalog shipment. In the kick-off meeting, the marketing and editorial departments brainstormed the themes of the next catalog. In the following weeks, first an 80 percent title list and then a 100 percent title list were compiled. The 100 percent book list was composed of all titles that were slated for inclusion in the next catalog. Consequently, it was a milestone in the process and had to be completed 34 days before catalog shipment. This was particularly important for the "new new titles" because the initial contract had to be negotiated with the publisher.[1] Once the layout was determined, catalog printing began. This was at a minimum of 10 days before catalog shipment and implied that all processes were terminated and no more changes regarding content or layout were possible at this point. This special deadline was important in contrast to previous deadlines after which many changes were still accepted and incorporated. These changes usually resulted from the editorial team's frequent last-minute additions of new titles to provide the "freshest" catalog possible—without taking into account the consequences of extending business deadlines.

Inbound Logistics

Inbound logistics were very similar for BBC and 21st Century. For the club business to be successful, not only did the catalog content have to be appealing, but also the proposed books needed to be physically available on time. This was the responsibility of the publishers that delivered their books directly to the two warehouses. Orders were supposed to arrive here at approximately the same time. The 100 percent title list was the trigger for inbound logistics, after which it took another two to four days before negotiations with the publishers were terminated and the contract signed. The contract maintained a fixed delivery date, usually around a week before catalog shipment, by which point the publishers were supposed to have brought the books to the central warehouses. In case publishers had more time to produce the order than actually necessary, they would proceed with another, more urgent, request. It usually took 20 to 30 days to deliver the ordered quantity of books.

Unfortunately, due to predrawing other clients' orders, Bertelsmann's timely orders were not always produced and delivered on time. A track report of the book deliveries showed that promised and actual arrival date differed on average by more than 10 days

(Exhibit 4). Meanwhile, the Chinese legal environment and BBC's and 21st Century's dependency on some large publishers for a large share of their merchandise made it difficult to enforce smoother contract handling. Therefore, the delivery time was virtually at the publishers' discretion.

Outbound Logistics

Outbound logistics for the two entities, in contrast, differed widely. BBC had to process mail orders for its club business and supply nine shops, all in close proximity. Its books were supposed to be in the warehouse on the catalog shipment date. However, a few days of delay were not very significant in the mail business because it took three to four more days before the members received the catalog and started ordering. Supplying its stores was relatively easy as they could all be reached within one day, and because no batching of orders was necessary, orders could be placed daily.

Meanwhile, 21st Century needed to arrange its supply chain so that it could supply all 28 shops outside of Shanghai on the catalog shipment date, but this proved to be very difficult in light of the long distances, the poor road conditions, and, especially during the winter, the harsh weather. (For an overview of 21st Century's distribution network consisting of a consolidation warehouse in Shanghai and 16 shipping lanes across China, refer to Exhibit 5.) As catalogs were present in shops on the shipment day and the shops were facing strong competition from other chains to provide new books as early as possible, books needed to be in 21st Century's warehouse much earlier than in BBC's. Many employees at BBC were unaware of 21st Century's growing importance for Bertelsmann's success in China and the difficulties they faced in shop business.

Regarding 21st Century, shops could only order once a week. Every Tuesday, the internal outbound logistics cycle started with the warehouse manager sending an inventory list to all shops. The shops were given two days to place their orders. 21st Century's 3PL started preparing the trucks the following day. The first trucks left the warehouse on Saturday. Although some shops were only located a day's drive away, they only received their orders on a Monday because the shops were too busy over the weekend to process incoming inventory. The outlying shops were situated a week's drive away and needed to wait until the following Friday. Hence, the whole process from receiving the order list to receiving the books in the shop took six to ten days.

Book Arrival Situation at Shop Level

The main problem for 21st Century was that books, especially the new books of a catalog, arrived in the shops too late—long after the catalog shipment day. Thus, important revenue was lost. Moreover, at the end of a book's life cycle, redundant books piled up, which could have been sold had they arrived earlier. An assessment of the situation at the shop level showed that one day before mailing the catalog, only 29 percent of the new new (NN) titles had arrived at shops located close to Shanghai, and at more distant shops, a mere 9 percent were available (Exhibit 6).

The delay was extremely critical for two reasons. First, books were comparable to fashion: They were most interesting to the customer when they were new. Most books had the highest sales during their first three weeks. Second, those NN titles that were late were in many cases the best sellers because the most successful titles were often the hardest to source. Nevertheless, Bertelsmann's competition seemed to be able to offer those titles earlier on several occasions. An analysis of the sales of the NN titles showed that the top 20 titles actually contributed 60 percent of the total sales of NN titles. In March, however, 10 of the top 20 titles arrived at the warehouse after the catalog shipment (Exhibit 7).

21st Century's problem with late arrivals already started at the consolidation warehouse. For example, only 75 percent of NN titles ordered had gone through the 21st Century warehouse by the catalog shipment date in March, one of the most sales-intensive months of the year. With lead time taken into account, this implied that even less stock was in the shops. The situation at the BBC warehouse was similarly troubling, yet slightly better (Exhibit 8).

Inventory Situation at Shop Level

Concerning the stock situation at the shops, three mutually dependent factors played an important role: the way in which the initial forecast was estimated, the reordering process, and the key-performance indicators of the shop managers.

The initial demand forecast had already been calculated by the editors of 21st Century. As mentioned already, forecasting the sales of media products was particularly difficult, and the accuracy of the forecasts of the editors of BBC and 21st Century was not monitored. Even though the editors were close to the market, in the end they had to rely on their intuition to come up with an initial forecast.

Shop managers had the full responsibility for reordering. It was assumed that the late arrival of books at shops drove the shop managers, believing they were undersupplied, to have a tendency to pad their orders. To avoid this, the 21st Century management prohibited returning books to the warehouse. This reflected the agreements that BBC and 21st Century had with their publishers, most of which did not allow returns either. The result was that, on an average, shops relied on three to five times more stock than needed (Exhibit 9).

There was no alignment between the key performance indicators (KPIs) used to derive the shop managers' bonuses and Bertelsmann's overall interest. As shop managers were employees of 21st Century, they did not become the owners of the books they ordered for their shops. Meanwhile, their KPIs mainly measured sales performance. Consequently, shop managers only bore the risk of under- but not overstocking, and thus had an incentive to pad their orders, many looking for creative ways to store books after they have reached their capacity limits.

Reality showed that it was virtually impossible to forecast the demand for books in a retail business and in particular for 21st Century. First, the shop business was a push supply chain where the customer order was only known at the last possible stage. Second, forecasting demand of media products such as books was complicated by the fact that they belonged to the group of fashion items where sales success depended to some extent on personal taste. Third, books often had a very short life cycle (three to six months). Consequently, initial decisions about order quantities were critical because the stores had only a few months to sell most of the stock. Finally, the high number of books launched every year presented another challenge. Every two months, more than 100 new books were offered in the Bertelsmann main catalog.

Foguang Chou decided to call his old friends at headquarters in Shanghai; it was time to work out a solution with his fellow store managers and the directors of the Book Club and 21st Century. There had to be a better way to handle orders and finally see books arrive on time. Something had to be done about the high levels of inventory and the chaotic ordering situation. He knew that he would face some difficult discussions during the upcoming meeting as it would be a challenge to avoid casting blame and ensuring that everyone saved face. At the same time, the chance to present the next new *Harry Potter* book to his customers on time—and possibly even before the competitors—made him smile.

Discussion Questions

1. Describe and visualize the supply chain at Bertelsmann Direct Group China. Consider both the Book Club and 21st Century. Map goods, funds, and information flows. (You can use Exhibit 10 as a template.) Then describe and visualize the business unit–specific outbound logistics cycle of 21st Century. Construct the process of the periodical catalog creation at Bertelsmann Direct Group China. Then compare it with 21st Century's in- and outbound logistics cycle. Where are the bottlenecks and why do books arrive late at the shop level? Provide suggestions for improvement in the process so that books can be guaranteed to arrive in the shops on time.

2. Describe the concept of the push-pull view of a supply chain by identifying and describing five generic push-pull boundaries. Into which category can the Book Club and 21st Century be categorized, respectively? How do you judge the strategic implications on the supply chain?

3. What role do cultural differences play in decision making at Bertelsmann Direct Group China? First, define the terms *organizational culture* and *corporate culture*. Then take both the Chinese culture as well as Bertelsmann AG's corporate culture into account and use the value, norms, artifacts, and patterns of behavior framework to explain any differences in detail.

4. Briefly describe the stock situation on a shop level for 21st Century. Name the most important drivers of the current inventory level and their direct consequences. What are the root causes of the high stock level in comparison with BBC? Do you see a connection to the delayed books? Is the incentive system for store managers appropriate? Provide suggestions for improvement.

5. Describe the concept of network design. What are its major drivers and what are the most important indicators of supply chain performance? Which indicators are particularly important at Bertelsmann Direct Group China and which are the corresponding supply chain drivers?

Bertelsmann China—Part B

It was one week before New Year's Eve 2006 and Liu Feng, a supply chain coordinator with Bertelsmann China, was sitting in the Shanghai headquarters. The noise from the outside was distracting and because nothing but smog seemed to enter the open windows, it was becoming almost impossible to concentrate.

Liu knew that time was of the essence—he had to present a detailed report to the vice president on how to lower the total logistics costs before taking the much-needed days off during New Year's. Liu Feng had thought about the problem many times before.

It had become clear to him that part of the solution was to extend the collaboration with the third-party logistics provider (3PL) and to add regional warehouses. He had not yet managed to solve the problem entirely, however, and tried hard to remember bits and pieces from his supply chain courses. It had been quite a while…

Key Insights from Part A

In Part A, Bertelsmann's Chinese book retail business and the corresponding challenges were introduced. Bertelsmann Group was a worldwide media and business services provider. Approximately 88,500 employees worked for the company's six divisions. These included the world's largest book publishing group (Random House, New York), Europe's leading magazine publisher (Gruner + Jahr, Hamburg), and the well-known Bertelsmann Music Group (BMG Music Publishing, New York).

Overview of Bertelsmann China

Bertelsmann entered the Chinese market in 1994. Two subsidiaries handled the book retail business. The Bertelsmann Book Club (BBC) operated shops in the Shanghai area and a catalog business. In 2006, Bertelsmann Book Club reached more than 1.5 million members with more than 7 million CDs, VCDs, DVDs, and computer games sold annually. BBC's management was based in Shanghai. Beijing-based 21st Century Book Chain (21st Century) ran 28 book shops, Bertelsmann Club Centers, in 16 cities all over China. (For an overview of the business units of Bertelsmann Direct Group China, please see Exhibit 5.)

Challenges for the Bertelsmann Supply Chain

In the summer of 2006, business performance was harmed by excessive inventories in retail shops. Books quickly became outdated and lost their value. Through the increase in stock, inventory holding costs were also rising. The key problem was books arriving late in shops, tempting shop managers to place excessive orders to assuage uncertainty.

Bertelsmann's Quest for Improvement

Bertelsmann tackled the challenges by improving supply chain management processes. The internal catalog development process was started two weeks earlier in order to ensure prompt book arrival in all shops in the short term. In addition, centralized forecasting allowed for better estimates of actual demand. This was crucial because the demand for books tended to be very volatile. It was difficult to know in advance whether an article would be a best seller or not sell at all. Furthermore, Bertelsmann expanded its incentive system by adding a key performance indicator based on stock levels in shops. Consequently, managers had an incentive to reduce inventory costs. In the end, the overall inventory situation improved substantially.

Insight into the Supply Chain Network of 21st Century

Besides the selection of the best-selling books and marketing, supply chain management was the most important success factor of Bertelsmann operations in China. The process was divided into two parts. First, the books were sourced from three major publishers and transported to the central warehouse in Shanghai. The books were then distributed to the 16 cities where Bertelsmann Book Club Centers were located (see Exhibit 5).

How Inbound Logistics Work at 21st Century

21st Century focused on three major publishers that accounted for roughly 85 percent of total sourcing volume. Local publishers were usually not considered when ordering new articles. Two publishers were based in the capital as China's major publishers were still located in geographic proximity to the government. They served a maximum of 4,000 and 5,000 units per order cycle, respectively. The third publisher operated from Shanghai and could provide as many as 7,000 units per order cycle. A logistics service provider transported the goods to the Shanghai warehouse and was paid according to a fixed rate per day of transportation. As 21st Century granted all shipments in the complete logistics process to this provider, it was charged only RMB (Renminbi) 0.61 per unit per day of transportation on all routes to and from the central warehouse as well as on other routes if required. Moreover, each day of shipment led to RMB 0.16 of inventory costs.

The Distribution Network in Detail

In the winter of 2006, 21st Century ran only one warehouse, which was located on the outskirts of Shanghai. All goods were shipped through this warehouse because freight capacity constraints precluded direct shipping. To avoid further losses, stock in the Shanghai warehouse should have been zero at the end of each order cycle. There was no special internal charge for handling goods at the Shanghai warehouse, as most of the work was performed by the logistics service provider who included the costs into the fixed charge. 21st Century's shops were located in 16 cities: Shanghai, Beijing, Tianjin, Qingdao, Dalian, Shenyang, Harbin, Ningbo, Hangzhou, Xuzhou, Chansha, Chengdu, Shenzen, Guangzhou, Kumming, and Chongqing (see Exhibit 11). If there was more than one shop, the truckload of goods was distributed to a logistics hub and then delivered to the shops in smaller vehicles. Otherwise, the products were delivered directly by one truck per town. Transportation costs between shops in one city were negligible compared with overall transportation costs. By pooling inventory in each city, stock-keeping costs and variance could be decreased. Although forecasting demand for a specific book was very difficult, total demand could be assumed to be relatively constant over the year, with a short peak before the Chinese New Year. The goal of Bertelsmann China's supply chain management was to avoid oversupply of stocks at all cost.

Optimizing the Network

Bertelsmann optimized its network in terms of cost with the help of linear programming. The connections between the warehouse and the shops or publishers were called nodes. The three publishers were supply nodes, the warehouse in Shanghai was a transshipment node, and the shops were demand nodes. Each node was assigned a specific cost. The important cost considered was total logistics cost. Bertelsmann calculated this cost as the sum of transportation cost and inventory holding cost.

Need for Further Improvement

Although 21st Century saw huge improvements in the summer of 2006, it became obvious that the current supply chain design was outdated. The network expanded at a speed in terms of books shipped, which had not been anticipated at its planning. Reaching its limits was just a question of time. Therefore, 21st Century discussed two options of improving the current design. First, relationship with the 3PL provider could be extended to contract logistics, and second, there was still the possibility of tapping the full potential of the current network by adding more routes.

How Logistics Service Providers Offer Value Added

Logistics is a fast-growing business in East Asia. China's major trade centers are all located along the east coast. However, in the last few years, the industrialization of central China has led to major increases in the transportation flow within the country.

As a result, the need for professional logistics services grew. European and Northern American logistics service providers noticed the opportunity to extend their businesses into the Chinese inland and began to offer a broad range of services. There were two main concepts in the logistics services business. First, freight forwarders concentrated on shipping. They owned the equipment, ships, and trucks, but offered few or no value-added services. Second, 3PL providers were "asset light." They handled all processes connected to logistics but outsourced the pure transportation business to freight forwarders. 3PLs offered many additional services, including contract logistics and warehouse management. They also managed more complex supply chain designs as "peddling routes." Using this concept, several destinations are served sequentially by one truck.

Bertelsmann established a long-term relationship with one of the leading European 3PLs to run its Chinese book supply chain. Due to the growing shop network, 21st Century increased in importance for this supplier. In the fall of 2006, the 3PL pushed into contract logistics and began to offer innovative warehousing services. The company provided space in its Beijing warehouse (Exhibit 11) and accomplished all related tasks for a fixed fee of RMB 0.2 per book. This charge would have to be added to the cost of handling all books in the contract warehouse.

Improving Network Design

Liu Feng contemplated several theoretically possible network designs, one of which seemed to be especially interesting. The status quo required keeping all stock at the central warehouse at Shanghai. Network efficiency could be improved, however, by linking some of the shops. The shop network of 21st Century included four megastores. These stores all had large sales floors, and were all located on the east coast in Shanghai, Tianjing, Quingdao, and in the capital, Beijing. Theoretically, these stores could serve as regional warehouses to replenish other stores in nearby cities (Exhibit 12).

In addition, Bertelsmann would gain the advantage of shorter lead times. On the one hand, the resulting gain in responsiveness had major advantages. It could significantly increase service levels, raising customer satisfaction, and it could increase forecast accuracy, reducing safety stock. On the other hand, sales performance could suffer from the additional stockkeeping tasks.

A Long Night

The last truck had already left on the snowy roads. As Liu Feng looked at the map, it became clear to him that some things had to be calculated first. This would be a long night. A new concept was desperately needed to develop the future direction of the cooperation with the 3PL provider. He opened his Excel file and began to think.

Discussion Questions

6. Describe the network design currently used at 21st Century. Put yourself in Liu Feng's position. What are the common network designs and the implication for supply chain costs? Regarding 21st Century's growth strategy, how could the current network design be operated to reduce total logistics cost? Consider the location of the publishers and the possibility of implementing new routes.

7. Map the current network design as a minimum cost problem and calculate total network costs through linear optimization. Which quantities do you order from the three publishers?

8. The 3PL of 21st Century offers Liu Feng the opportunity to expand the contract logistics cooperation. 21st Century can rent a part of the 3PL's warehouse in Beijing. Does it make sense for 21st Century to use a second warehouse? Map the new situation.

9. Liu Feng considers having some shops supply other shops. Thus, the transportation routes of the trucks could be linked to all other shops. Liu Feng thinks that the shops in Shanghai, Beijing, Tianjing, and Qingdao are the most appropriate because they have the most floor space. What would 21st Century's transportation network look like with these "mega cities" and the second warehouse? Do you think Liu Feng's idea is feasible?

Endnotes

1. "New new titles," written as NN titles, are those titles presented in a BBC catalog for the first time. The phrase is influenced by Chinese grammar and is used like this in daily business.

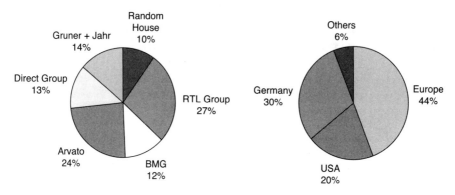

Data Source: Bertelsmann Direct Group China (2006)

Exhibit 1   Revenues of Bertelsmann AG by business unit and geographical area. Total revenue (2005): EUR 17.9 bn.

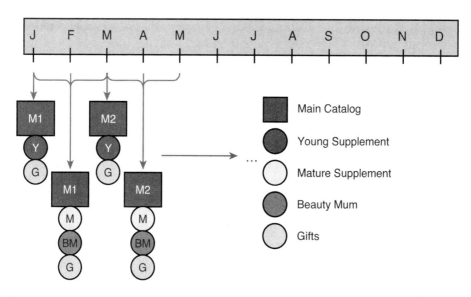

Data Source: Bertelsmann Direct Group China (2006)

Exhibit 2   Catalog sending process within Bertelsmann Book Club

	March						April			May
	0 days	14 days	20 days	30 days	31 days	32 days	46 days	49 days	55 days	65 days
	↓	↓	↓	↓	↓	↓	↓	↓	↓	↓
	A	B	C	D	E	F	G	H	I	J
	Kick-off	Title planning	80% title list	Trial balance	100% title list	Brief layout	Layout	1st draft	Catalog deadline	Catalog sending

Inbound logistics process of 21st Century triggered approx. 34 days before catalog sending date.

Data Source: Bertelsmann Direct Group China (2006)

Exhibit 3 Overall catalog development process timeline

ITEM	Name	Estimate Arrival Date	Actual Arrival Date	Difference
…	…	…	…	…
85469	阁楼上的光	23.03.2006	10.04.2006	18
85392	蓝石头	23.03.2006	07.04.2006	15
85414	蓝,另一种蓝	23.03.2006	07.04.2006	15
85413	迷失的男孩	23.03.2006	07.04.2006	15
85405	英语广场2005下半年合订本	23.03.2006	07.04.2006	15
85409	科学的答案我知道*C	23.03.2006	04.04.2006	12
85410	健康的答案我知道*c	23.03.2006	04.04.2006	12
85567	糊涂学	23.03.2006	04.04.2006	12
85386	古典今看-从诸葛亮到潘金莲	23.03.2006	03.04.2006	11
85486	老徐的博客	23.03.2006	28.03.2006	5
85393	人骨手镯	23.03.2006	28.03.2006	5
85406	电脑医院2006	23.03.2006	28.03.2006	5
85475	简奥斯丁全集	23.03.2006	28.03.2006	5
85395	陷阱	23.03.2006	28.03.2006	5
85396	鹅鹏案卷	23.03.2006	28.03.2006	5
85421	口才改变人生	23.03.2006	28.03.2006	5
85403	天下衙门	23.03.2006	28.03.2006	5
85470	健康蔬菜	23.03.2006	28.03.2006	5
85471	二人开伙	23.03.2006	28.03.2006	5
85424	一字禅	23.03.2006	27.03.2006	4
85407	西西里的传说*c	23.03.2006	23.03.2006	0
85408	五月之诗*c	23.03.2006	23.03.2006	0
85495	死神首曲(下)	23.03.2006	23.03.2006	0
85498	过把瘾就死	23.03.2006	23.03.2006	0
85497	看上去很美	23.03.2006	23.03.2006	0
85412	雨天的海豚们	23.03.2006	22.03.2006	-1
85399	扑克牌魔术	23.03.2006	22.03.2006	-1
85491	感悟(绿版)	23.03.2006	21.03.2006	-2
…	…	…	…	…

Mean (difference between estimated and actual arrival): 10.44 days
Data Source: Bertelsmann Direct Group China (2006)

Exhibit 4 Inbound logistics estimates of publishers are unreliable

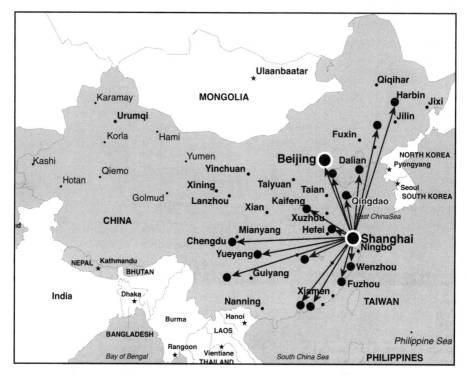

Data Source: Bertelsmann Direct Group China (2006)

Exhibit 5   21st Century distribution network design

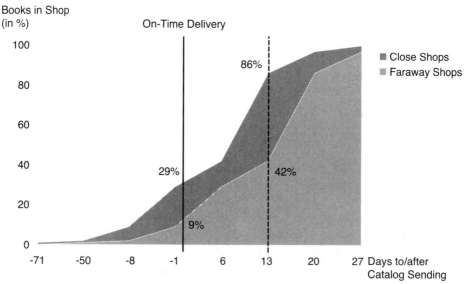

Data Source: Bertelsmann Direct Group China (2006)

Exhibit 6   Arrival situation of new books at the shop level

ITEM	March NN Ranking	Item Name	% of NN QTY sold	Days late at WH	Editorial Star Rating
84343	1	莲花*R	11,4%	13	★★★★★
84955	2	小王子经典珍藏版	7,0%	-7	★★★
84970	3	泡沫之夏	4,5%	12	★★★
84907	4	死亡拼图*C	4,2%	-2	★★★★
84522	5	甜心涩女郎	3,8%	13	★★★★★
84926	6	面包树上的女孩	3,4%	1	★★★★
84916	7	韦特塔罗精装修订版	3,4%	0	★★
84924	8	伤心致死	2,5%	7	★★★
84036	9	印记	2,3%	13	★★★
84954	10	风度	2,3%	-7	★★
84923	11	我叫金三顺	2,2%	7	★★★
84981	12	读者文摘2005合订(秋季+冬季)	2,2%	-12	★★★
85044	13	遗忘爱	2,1%	6	★★★
84925	14	卡耐基黄金50年	2,0%	-15	★★
84929	15	食物是最好的药2	1,8%	-7	★★★
84935	16	我的可爱老师	1,8%	6	★★
84968	17	悟空传*R	1,8%	-21	★★
84919	18	瓦尔登湖	1,5%	-1	★★
85004	19	我的人生笔记	1,5%	-9	★★★★
84971	20	二战美军战列舰.巡洋舰	1,4%	-15	★★
			63,1%		

Dark: Late at the warehouse
Medium: Late at shops
Light: On-time on catalog sending day
Data Source: Bertelsmann Direct Group China (2006)

Exhibit 7   Late arrival of top ten books at the warehouse

Book arrival in warehouse by catalog sending day	March	April	May
BBC	76%	89%	98%*
21st Century	75%	76%	77%

* Measured on May 7, not the first day of the month
Data Source: Bertelsmann Direct Group China (2006)

Exhibit 8   Arrival situation at the warehouses

Category	Monthly sales (units)	Shops	Avg. overstock factor*		Target overstock factor (at Q*)	
A (6 shops)	150,000 - 310,000	Hangzhou, Beijing (3), Shanghai**, Shenzen	April 4.16 May 5.33 June 6.92	5.47	April 1.09 May 1.05 June 1.03	1.06
B (11 shops)	70,000 - 149,999	Ningbo, Dalian (2), Xuzhou, Guangzhou, Tianjin (2), Beijing (2), Qingdao, Chengdu	April 4.22 May 3.91 June 3.36	3.83	April 1.04 May 1.07 June 1.03	1.05
C (11 shops)	<70,000	Chongqing, Kumming, Harbin (2), Shenzen (2), Beijing, Shenyang (3), Chansha	April 3.78 May 3.18 June 3.06	3.34	April 1.03 May 1.03 June 1.02	1.03

* The overstock factor is defined as the factor by which the actual stock level is larger than an optimal stock level computed under the "newsboy model."
** Outside of Shanghai city limits.
Data Source: Bertelsmann Direct Group China (2006)

Usage of the "newsboy model" to determine the optimal stock level under uncertainty:
Determine Q*, the optimal stock level, where

$Q^* = \mu + z * \sigma$ with $z \to a = c_u / (c_o + c_u)$

- Q*—Optimal stock level
- μ—Average demand
- σ—Standard deviation of demand
- z—Statistical value from the normal distribution, derived from a
- a—Optimal service level
- c_u—Underage costs
- c_o—Overage costs

As a result, the optimal service level for 21st Century shops is around 67% because of high overage costs. The corresponding inventory levels were exceeded by a factor of 3.3 to 5.5.

Exhibit 9   Stock situation at the shop level

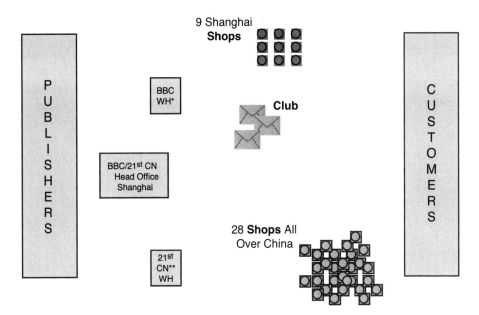

* Warehouse
** 21st Century
Data Source: Bertelsmann Direct Group China (2006)

Exhibit 10 Template for mapping the Direct Group China supply chain

Lead-times (days)		
From	To	
	Shanghai WH	Beijing WH
Beijing P1	3.19	0.70
Beijing P2	3.51	0.90
Shanghai P	0.85	3.19

To	From		Demand
	Shanghai WH	Beijing WH	
Shanghai	1.06	3.19	1700
Beijing	3.19	1.06	2300
Tianjin	3.08	1.08	550
Qingdao	3.19	2.12	650
Dalian	4.30	2.05	800
Shenyang	4.25	2.12	400
Harbin	5.31	3.19	900
Hangzhou	1.06	4.25	300
Ningbo	1.09	4.27	300
Xuzhou	2.12	2.12	650
Chansha	3.19	4.25	750
Chengdu	7.44	6.38	350
Shenzen	4.25	7.44	1100
Guangzhou	4.23	8.51	400
Kumming	5.31	7.44	650
Chongqing	7.56	8.07	500

Data Source: Bertelsmann Direct Group China (2006)

Exhibit 11 Lead times according to the contract with the 3PL and average demand levels per order cycle in 2006

	Lead-time (days)				
	To	From			
		Shanghai	Beijing	Tianjing	Qingdao
1	Shanghai	0.00	4.14	3.92	4.43
2	Beijing	4.14	0.00	1.38	2.76
3	Tianjin	3.92	1.38	0.00	1.29
4	Qingdao	4.43	2.76	1.29	0.00
5	Dalian	5.52	2.83	2.59	3.24
6	Shenyang	5.46	2.76	3.89	5.19
7	Harbin	6.90	4.14	4.54	7.25
8	Hangzhou	1.38	5.52	1.29	0.64
9	Ningbo	1.32	5.52	2.59	3.89
10	Xuzhou	2.76	2.76	5.22	3.86
11	Chansha	4.14	5.52	5.19	5.19
12	Chengdu	9.67	8.28	4.54	5.31
13	Shenzen	5.52	9.67	9.09	7.79
14	Guangzhou	5.54	11.05	9.74	8.66
15	Kumming	6.90	9.67	9.09	9.23
16	Chongqing	9.67	9.80	6.38	6.49

Data Source: Bertelsmann Direct Group China (2006)

Exhibit 12 Lead times in case of a shop-to-shop delivery

6

CARNIVAL CORPORATION FOOD SUPPLY CHAIN

Dick Verbeek, CPIM CIRM CSCP, Supply Chain Skills

Introduction

Luigi Giordano, director of food operations, anxiously opened an e-mail attachment from corporate headquarters. The memo from his new boss, Reginald Cooper, senior vice president of operations, read as follows:

MEMO TO ALL NORTH AMERICAN DIRECTORS OF FOOD OPERATIONS

Your management team is committed to ensure that the corporation has strategic plans in place that will meet the needs of its shareholders and customers. In support of this mandate, the Senior Management Team will begin a review and update of the corporation's long-term strategic plans next month.

Our market research indicates that the North American cruise industry will continue to enjoy significant growth for several decades. We also expect significant competitive initiatives designed to attack our dominant market share position. These initiatives will likely include price discounts from some of our major competitors. We have identified the food supply chain as one component of costs that has significant opportunities for efficiency gains.

During the next week, please conduct a thorough review of your ship's food supply chain to identify cost-cutting opportunities. Our overall corporate objective is a 20 percent reduction in food costs on a per-passenger basis. Your contribution toward this goal is critical to the long-term success of the corporation.

I look forward to reviewing your detailed recommendations.

Sincerely,
Reginald Cooper

Giordano took a deep swallow of his morning espresso. He knew there were going to be some long nights ahead!

Carnival Corporation Overview

Carnival Corporation & plc is the largest global cruise line operator and one of the largest vacation companies in the world. The company operates under several well-recognized brands, including the following:

- Carnival Cruise Line
- Princess Cruises
- Holland America Line
- Cunard Line
- Costa Cruises
- P & O Cruises

Carnival is headquartered in Miami, Florida, USA, and London, England. It operates a fleet of more than 80 ships and typically has more than 150,000 guests and 65,000 shipboard employees sailing at any given time.

Table 1 summarizes key financial data for the corporation. (All figures are given in millions.)

North American Cruise Industry

Giordano's ship has been assigned to service the North American market. It performs Caribbean cruises during the winter season, which extends from November through April. A typical Caribbean itinerary would be seven days with three to four ports of call that could include the Bahamas, U.S. Virgin Islands, Puerto Rico, and the Turks and Caicos. Departures are typically from the Port of Miami.

For the summer season, June through August, the ship is repositioned to service the North East coast. Departing from New York, the four-day itinerary usually includes a port stop in St. John, New Brunswick, Canada.

The North American cruise industry is dominated by three major organizations. In addition to Carnival, Royal Caribbean and Norwegian Cruise Line combine to control more than 80 percent of the market.

The North American market has experienced steady growth in demand, as summarized in Exhibit 1.

Giordano's Thought Process

Giordano, a 10-year "veteran" in the cruise industry, decided that a rigorous fact-driven analysis would be the best approach in the formulation of recommendations for the head office. He decided to involve the ship's head chef to assist in the identification of issues and the generation of potential solutions. He also felt that key representatives from some of his major suppliers could provide valuable insights. To facilitate everyone's involvement, he proceeded to call a private meeting with head chef Timothy Rousseau and immediately dispatched e-mails to his contacts from two major suppliers based in Miami and New York, respectively.

Meeting Notes with Head Chef Rousseau

Chef Rousseau entered Giordano's cabin in an authoritative fashion, slamming the door and immediately pronouncing, "Don't those idiots in the head office know that you can't cut corners when it comes to food! Our guests demand the best and they deserve the best!"

"Ease up, my friend," Giordano replied. "You have misunderstood our mandate. The corporation has asked us to maintain the same variety and high standards that our passengers expect from us. However, they are asking us to closely examine our food procurement, storage, and preparation to pinpoint any opportunities to reduce waste and costs. Timothy, I would like you to concentrate on generating as many ideas as possible to reduce the cost associated with food preparation. One area that continually irritates me is the amount of food we throw out. Did you know that we throw out approximately $30.00 of food per passenger every cruise?"

Chef Rousseau assumed a defensive posture and replied, "This is a cruise ship, Luigi. People expect variety at the buffet! It's impossible to predict the number of people who will eat at the buffet on any given day!!" On that note, Chef Rousseau stormed out of Giordano's office.

Giordano was concerned about Chef Rousseau's negative reaction. He felt that it could seriously hamper his ability to identify tactics to reduce food waste.

Typical Food Consumption

Giordano continued to collect data from corporate databases. After considerable searching, he was successful in the identification of a food consumption database for his ship. From ad hoc queries, he was able to develop the food consumption profile shown in Table 2.

Given the existence of this data, he scratched his head for a few moments and asked himself, "Why do we have such high spoilage rates? Shouldn't we be able to predict our weekly needs with greater precision?" Giordano noticed a menu item on his computer screen titled "Predicting Onboard Material Requirements" when his pager suddenly alerted him of an emergency in the engine room.

Food Consumption Variables

Predicting food consumption for more than 4,000 people—3,000 guests and 1,000 crew—is more difficult than expected due to a variety of factors. Chef Rousseau summarized the following issues in his written response to the director:

- Each week has 3,000 *new guests* with different tastes, food allergies, and eating habits.
- The total number of guests can fluctuate from week to week.
- Some guests routinely participate in shore excursions when visiting a port of call. These passengers eat less food onboard the ship for the day. Other guests prefer to remain onboard at some or all ports. The percentage of "excursion-prone" guests versus "homebodies" varies from sailing to sailing.
- Occasionally, the itinerary is changed due to weather conditions or mechanical problems. When this happens, the galleys experience higher food demands as no one is eating at the port of call.

Carnival Cruise Food Operating Policies

Quality control is critical for all cruise lines. As part of the quality control procedures at Carnival, the ship is restricted to only procure and prepare food and beverages that are FDA-approved. Effectively, this eliminates the ability to purchase extra supplies at various ports of call.

Additionally, the ship is mandated to carry two weeks of stock when leaving the Port of Miami. This safety stock of food provides "insurance" should the ship become disabled or stranded at sea for an extended period of time.

The Food Procurement Process

Currently, the procurement of food is a very informal and rushed process. The basic steps involved are summarized in Table 3.

Giordano utilizes a simple consumption replacement model for most food items. For example, he targets 96,000 eggs prior to leaving Miami. When he places his egg order, he orders sufficient eggs to "top up" his remaining stock to the 96,000 target.

Suppliers are specially chosen for their ability to deliver high-quality goods in the very short six-hour time frame that the ship is anchored in Miami.

Giordano feels that the replenishment process, while easy to understand and operate, may need a major redesign to satisfy the objectives established by his new boss. He asked his two suppliers to assess the process and provide ideas for improvement based upon their interaction with competitors and other industries. He specifically requested that they "identify procurement best practices" in his e-mail to his supply chain partners.

In addition, he has always felt that he should be taking advantage of the company's new customer preferences database. The database, part of a larger marketing analytics system purchased from SAS earlier that year, allows Giordano to predict customer preferences for everything available onboard. This includes excursions, spa treatments, and, most important, food and beverage biases. Unfortunately, Giordano has been too busy to participate in the online corporate training program that would teach him the skills required to operate the new system.

Food Staff

520 of the 1,000 crew members are involved in the preparation and delivery of food onboard the ship, as shown in Table 4.

Carnival offers competitive salaries to all of its crew members. Cooks are paid on a sliding salary scale that rewards their job position and years of service. In addition, they are encouraged to participate in online culinary academy courses to improve their skills. There is high employee turnover in the other food operations positions due mainly to long hours of work and low salaries.

Carnival's Food Supplier Base

Carnival deals primarily with a number of full-service distributors. These distributors work with a number of brokers and manufacturers to offer a variety of food and beverage items to Carnival. They also coordinate all logistical details affecting delivery to the port of debarkation. Prior to being placed on the qualified supplier list, a distributor must meet minimum requirements. These are listed in Table 5. Pricing for all products is negotiated by head office buyers with the appropriate distributor representatives and is typically formalized with a 6- to 12-month fixed-pricing contract. More than 80 percent of Carnival's food purchases are covered by fixed-price supplier contracts.

Approximately 95 percent of Carnival's needs are received at the port of debarkation. The remaining volume is replenished at a port of call and is typically reserved for "topping up" fresh produce.

Companies that meet the minimum requirements are assigned a score based on their relative total costs of ownership, on-time delivery performance, and selection. The top five distributors/suppliers are then placed on the preferred supplier list. As a general guideline, each ship is asked to spend 80 percent or more of its food budget across the top three on the list. Separate lists exist for packaged goods, fresh produce, and meats.

Finally, each supplier generates a product list that meets or exceeds Carnival's requirements for brand recognition, longevity (i.e., ability to stay fresh for one week), price, quality, lack of packaging, and ease of preparation. Minimal packaging is critical because waste and garbage must be stored onboard for the duration of the trip. Ease of preparation is important because of the large turnover in the cooking staff. Ideally there is little to no training time for a new cook to work with a product. The ship's staff can order any item in the quantities it requires from the distributor's product list. Occasionally, head office buyers will work with the distributor to create a Web site of "Carnival Hot Deals." These are typically promotional deals of limited time availability. Ship personnel, like Giordano, are allowed and encouraged to order food and beverages that best meet the needs of their passengers for the coming voyage.

The two suppliers selected by Giordano for specific feedback are his preferred distributors for packaged goods and meats. They represent 95 percent and 90 percent of his ship's annual purchases for packaged goods and meats, respectively. EatWell Distributors is also a preferred supplier of fresh produce. Both organizations have two-year contracts with Carnival. Pricing terms with both are net 45 days.

Competitive Benchmark Data

Giordano felt that the 20 percent reduction objective may have been set arbitrarily. Therefore, he decided to download key competitor cost and passenger benchmarks for Royal Caribbean Cruise Lines and Norwegian Cruise Lines (NCL).

Table 6 summarizes 200z–200x revenues and expenses for Royal Caribbean Cruise Lines. (All values are given in thousands.)

When Giordano attempted to access financial data for NCL, he discovered that his competitor's financial data was combined with the other operations owned by parent company Star Cruise Lines. With no obvious method to separate NCL's North American data from Star's worldwide corporate data, he decided to omit NCL financial data from his benchmarking analysis.

Passenger data for Carnival and Royal Caribbean are summarized in Table 7.

Giordano noticed that Royal Caribbean also analyzed expenses as a percent of revenue. He decided to perform the same analysis for Carnival. His analysis of food costs as a percentage of total revenues is summarized in Table 8.

Supplier Recommendations

Giordano received e-mail responses from his two supplier contacts. Their responses are reproduced in the following sections.

Memo from Mark Bell, Customer Support Coordinator, EatWell Food Distributors

Luigi,

Thank you for your e-mail message. I believe that my company can help you achieve your corporate objectives for cost reductions. Specifically, I recommend that we explore:

1. Buying more staples and dry goods in bulk. We can hold the purchased food in our storage facilities on a consignment basis while you draw from it on a weekly basis
2. Switching some items to less-expensive suppliers.

I should also point out that my organization has developed a new suite of software tools to control food consumption and plan replenishment. Our Web site has complete details, including a demonstration package available for download. Some of the major highlights are contained in the attachment to this e-mail.

At your next stop in Miami, I suggest we have a meeting to review specifics and to provide a complete software demonstration.

Contents of File Attachment

ANNOUNCING EATWELL's FOOD AND BEVERAGE INVENTORY MANAGEMENT SUITE

There are three modules provided for use on board the vessels:

- **EWICS (Inventory Control System)**—This module is used for inventory control, requisitions, ordering, and receiving of your food needs. The module includes a vast array of reports and leading-edge performance metrics related to inventory, consumption, and orders. Metrics include inventory turns and cash-to-cash conversion cycle.

- **EWFBS (Food and Beverage System)**—This module is used for tracking of beverage sales from the Point of Sale (POS) system. It includes detailed reports related to meal counts and beverage popularity.

- **EWMCS (Meal Count System)**—This module is for tracking meal orders placed during service time. The system operates on a notebook computer in the galley. The system provides an up-to-the-minute summary of dining room orders. The chef can access this information at any time during dinner service to determine if additional quantities of certain dishes will be required. The data can be exported on a daily basis into the EWFBS system.

All three systems are designed to operate on any PC operating Microsoft Windows.

Memo from Lucy Lawrence, Sales and Client Care, Analyst L&M Meats

Luigi,

I am glad you contacted me for suggestions to address your corporate cost-cutting initiatives. The best advice that I can give you is to maximize the flexibility in your supply process. For example, most of my other customers take advantage of the multiple distribution points we have in North America. Did you know that we have meat distribution facilities in Halifax, Nova Scotia, and Puerto Rico?

In addition, it would help me greatly to have a forecast of your food needs. If you can supply a weekly food forecast, I can be more efficient. In turn, I can pass on some of the savings to you.

Lastly, I believe there is a *huge* opportunity to reduce the average value of your inventory by carrying significantly less inventory of the expensive "dinner meats" (lamb, veal, etc.) and leveraging our Puerto Rico facility. Here are the high-level details:

1. Reducing the amount of "dinner meats" from 14 days of supply (DOS) to five DOS when you depart Miami.

2. As your Caribbean itineraries always stop in Puerto Rico, we can ship you two DOS of fresh dinner meats halfway through your cruise.

3. To address your corporate policy of setting sail with two weeks of food, I propose increasing your DOS of packaged "breakfast" meats from 14 to 23 days.

I have performed a preliminary analysis of your current state in the attached spreadsheet.

You will note that the key to the plan is the relative value of each category of meats (Dinner $13K/DOS, Breakfast $3K/DOS).

Let me know when you have some time to discuss details over a cappuccino.

Attachment from L&M Meats

The contents of the attachment from L&M Meats are shown in Table 9.

Giordano's Recommendations

Giordano was one day away from the due date set by his boss for submission of his recommendations. He knew this was going to be a long evening as he still had many outstanding questions that required his attention.

1. Given food costs are virtually identical (see Table 8), is a 20 percent reduction in food costs a reasonable target for cost reductions? Is there a better way to measure performance other than food cost as a percentage of revenue?

2. What changes should be made to the food procurement process to reduce the material costs of the supply chain?

3. How can Giordano reduce the impact of food consumption variability?

4. What metrics should Giordano use to substantiate his recommendations to management?

5. How can Chef Rousseau contribute to the 20 percent cost-reduction target?

6. Which supplier recommendations should he include in his report?

7. What factors could explain the difference in performance between Royal Caribbean and Carnival when comparing food cost/passenger?

Table 1 Carnival Corporation Select Financial Data

Partial Consolidated Statement of Operations			
Revenues	*200x*	*200y*	*200z*
Cruise passenger tickets	$8,903	$8,399	$7,357
Onboard sales	2,514	2,338	2,070
Other revenue	422	357	300
	11,839	11,094	9,727
Costs and Expenses			
Operating			
Commissions, transportation	1,749	1,645	1,572
Onboard and other	453	412	359
Payroll and related	1,158	1,122	1,003
Fuel	935	707	493
Food	644	613	550
Other ship operating	1,538	1,465	1,315
Other noncruise	314	254	210
Total	6,791	6,218	5,502

Data Source: Carnival Corporation 2009 Annual Report

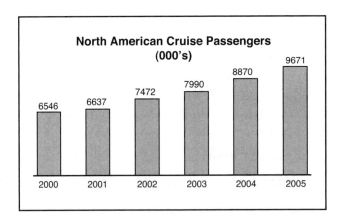

Data Source: CLIA Cruise Lines International Association

Exhibit 1 North American cruise passenger growth

Table 2 Weekly Food Consumption Profile

Food Detail	Value
Average value of weekly food purchases	$360,000
Average value of weekly food consumption	$230,000
Average value of weekly food waste (i.e., spoilage)	$130,000
Food waste:	
Overpreparation	55%
Spoilage (expiry dating)	35%
Other miscellaneous	10%
Average weekly meat (i.e., chicken, beef) consumption (in pounds)	20,000
Average weekly egg consumption	48,000
Average weekly milk consumption (in gallons)	1,000

Data Source: Carnival Corporation 2006 Annual Report

Table 3 Weekly Food Replenishment Process

Responsible	Timing	Task
Giordano	Two nights before the next cruise departure	Calculates total consumption of all items on board the ship
Giordano and Rousseau	One night before the next cruise departure	Discuss menu adjustments and plans for the next sailing
Giordano		Calculates food requirements for the next trip
Giordano		Reviews key supplier Web sites for specials, new items, etc.
		Places order with suppliers
Suppliers	Between 8:00 a.m. and 2:00 p.m.	Deliver all ordered supplies

Table 4 Food Operations Staff Data

Job Category	Number
Cooks	110
Dining room personnel	250
Utilities and support staff	160

Table 5 Minimum Supplier Requirements

Requirement	Details
Quality and safety	FDA-approved, compliant with Hazard Analysis and Critical Control Point (HACCP)
Accessibility and security	Must have operations at all ports of debarkation. Must be enrolled in Customs-Trade Partnership Against Terrorism (C-TPAT)
	Must be compliant with the requirements of the Bioterrorism Act (BTA) and the Country-of-Origin Labeling (COOL)

Table 6 Royal Caribbean Cruise Lines Select Financial Data

Partial Consolidated Statement of Operations			
Revenues	*200x*	*200y*	*200z*
Cruise passenger tickets	$3,838,648	$3,609,487	$3,359,201
Onboard sales and other revenue	1,390,936	1,293,687	1,196,174
	5,229,584	4,903,174	4,555,375
Costs and Expenses			
Operating			
Commissions, transportation	917,929	858,606	822,206
Onboard and other	331,218	308,611	300,717
Payroll and related	501,874	510,692	487,633
Fuel	480,187	367,864	251,886
Food	278,604	270,674	269,436
Other ship operating	739,817	677,785	687,505
Other			
Total	3,249,629	2,994,232	2,819,383

Data Source: 2006 Financial Report

Table 7 Passengers Carried

Company	200x	200y	200z
Carnival	7,008,000	6,848,000	6,306,000
Royal Caribbean	3,600,807	3,476,287	3,405,227

Data Sources: 2006 Annual Reports and Carnival Corporation, Royal Caribbean Cruises Limited

Table 8 Food Cost as a Percentage of Total Revenue

Company	200x	200y	200z
Carnival	5.4%	5.5%	5.7%
Royal Caribbean	5.3%	5.5%	5.9%

Table 9 Supplier Recommendations—L&M Meats

Current State			
	Average Weekly Purchases	*DOS*	*Price/DOS*
Dinner meats	$95,000.00	7	$13,571.43
Breakfast meats	$25,000.00	7	$3,571.43

Leaving Miami:			
		DOS	*Value*
Dinner		14	$190,000.00
Breakfast		14	$50,000.00
Total			$240,000.00

7

DSM MANUFACTURING: WHEN NETWORK ANALYSIS MEETS BUSINESS REALITY

John R. Macdonald, Michigan State University
Kelvin Sakai, UTi Worldwide Group

Introduction

Jim and Phil stood at the rental car counter after arriving from separate coasts for their third and final meeting at DSM (Des Moines) Manufacturing, a *FORTUNE* 500 and global consumer products company. As external consultants, they had been hired to work on a supply chain network analysis for DSM a few months earlier, and while the analysis itself had gone well, they were struggling with their final recommendations. "I'll chip in $30 extra if we can get the Mustang today," Jim said to Phil. "I need the extra muscle and energy to think through all these meeting details." Phil smiled knowingly and requested the Mustang.

Thirty minutes later, as Phil was weaving in and out of traffic on the way to Marshalltown, Iowa, Jim sighed for the third time, and started to reflect upon how a seemingly straightforward network analysis had turned into something much more.

Typically, Jim thought, network analyses quantified the cost savings and/or service-level improvements associated with proposed changes to an existing supply chain. Examples of such changes might include opening or closing a warehouse, shipping goods direct, or perhaps utilizing trucks more effectively by filling them closer to capacity. What happens, though, when the existing supply chain is relatively inflexible? Where are the improvements supposed to come from? And what about those other important aspects of an efficient supply chain that a network analysis does not necessarily address? Should you just ignore these questions as if they do not exist? All these uncertainties were troubling Jim as their car made its way to Marshalltown.

History

DSM recently celebrated its 100th year in business; it had been around a long time and withstood many changes. One hundred years ago, Iowa's population represented a greater proportion of the general U.S. population than compared with that of today.

Its customer base in the United States was centrally located in the middle of the country between Chicago, Minneapolis, Omaha, Kansas City, and St. Louis. However, times had changed and the company's traditional domestic customer base had migrated to the East and West Coasts of the United States. At the same time, DSM had aggressively become a national, and later a fully global company, selling its many products at home and abroad. In the past 30 years, DSM had been slowly outsourcing many of its core supplier activities to other U.S. locations outside of Iowa, then to Mexico and Europe, and, later, to Asia, with large plants in China, Thailand, and Malaysia.

Many of the overseas raw material parts came in through the Midwest gateway of Chicago via the West Coast and then backtracked to Marshalltown. This was because DSM was committed to the community of Marshalltown, where it continued to maintain its primary final production plant, as well as its engineering and research and development center. As a result, virtually all DSM products, including the one at the heart of Phil and Jim's current project, had to pass through Marshalltown to complete production.

Project Description

The initial description of the project had been detailed to Phil and Jim the first time they went to Marshalltown for meetings. More details and complications had emerged, however, over the course of the project.

The scope of the project was to focus on the Home Paint Products Division and the various global suppliers that shipped goods to the United States for that division. These suppliers shipped primarily via ocean. This scope is illustrated in Exhibit 1, with the corresponding ocean rates provided in Table 1. For the purposes of the analysis, transportation costs were assumed to begin from the origin port of the supplier. Air shipments were used only on an emergency, as-needed basis.

Examples of products from the Home Paint Products Division include special nontoxic paints designed for children's rooms, with various colors, brushes, industrial paints, and others. These goods would come into the United States and then be shipped by rail and/or truck to either Marshalltown (for random quality-control testing), or a designated contract manufacturer in Iowa or Minnesota. A contract manufacturer in some instances would take the product as a raw material input for another manufactured product, or could act more as a copacker and simply match products from different sources, package them together, and ship them to a regional distribution center (RDC) for later sale to a retail customer.

The new product driving the DSM supply chain review, called "Project Fragrance," was a paint that slowly released fragrance over the course of its first year of life. Project Fragrance came with a single-use can of paint, which was to be packaged along with a paint brush. The brush came from China while the paint was produced by the primary Marshalltown manufacturing plant. The Marshalltown plant had a local Marshalltown supplier provide the paint cans for filling. The contract manufacturer would then package (shrink-wrap) the brush and paint together to form a single item that would receive a printed label next and be shipped to an RDC for later sale.

Because it was a new product for the division, Project Fragrance was the primary focus of the network analysis. The initial 10 percent of imported volume (brushes in this case) from Asia, and a random 10 percent ongoing, were required to be sent first to Marshalltown for quality inspections before being sent on to the contract manufacturer for final packaging.

After a brainstorming session in the second meeting, Jim and Phil were asked to identify the best location for a single contract manufacturer that would result in the lowest total transportation costs for DSM. The supplier and contract manufacturing locations and predicted volumes are presented in Table 2. In addition, they were also asked to make any other observations and recommendations related to the division's supply chain network, which were discovered over the course of their efforts on Project Fragrance.

The Meetings

Gustavo Cortes was the transportation manager for DSM Manufacturing North America. He was the primary sponsor for the network analysis project and had led the first meeting. His assistant, Julie Rhodes, would be Jim's and Phil's primary contact for any data or information needed to model the DSM network. Jim and Phil would analyze the network using a modeling tool called ModelPro 2020.

"Instead of modeling the end customer," Gustavo said, "we have five RDCs that can serve as final points for the flow of product in your model. Those five locations are in San Francisco, Dallas, Atlanta, New York, and, of course, here in Marshalltown. The goal of the project is to minimize transportation costs both internationally and within the United States. We currently use a mixture of rail and truck to import the goods and move them between the coasts and plants to contract manufacturers and then on to the RDCs."

The five RDC locations, their service areas, and their relationships to existing contract manufacturing locations are illustrated in Exhibit 2. The relative size (in unit volume) of each RDC varies considerably from region to region. RDC volumes (as a percentage of total unit volume) are as follows:

- Atlanta—20 percent (serves Southeast)
- Marshalltown—35 percent (serves Midwest and the Rocky Mountain states)
- Dallas—12.5 percent (serves Mid-South up to Arizona)
- New York—20 percent (serves New England and the Mid-Atlantic region)
- San Francisco—12.5 percent (serves the West Coast)

Phil reconfirmed that rail was used to ship goods to the RDCs. Rail rates are provided in Table 3. Usually, service levels for getting goods into the customers' hands meant that trucks needed to be used for quicker service. Phil tried to dive deeper into the reasons that service levels were not important to DSM's customers. Julie interrupted and said, "Well, the real issue that we have is our inventory. We can't see it! Right now, the contract manufacturers tell us only when we call them how much product they have. Of course, they bill us for a number of units each month, but it is frequently different than what we consider a reasonable amount. I personally think that we should stop outsourcing anyway. My transportation costs would go down if we could bring this simple repackaging back to Marshalltown instead of sending trucks 150 miles to some of the contract manufacturers. I hope your analysis shows Marshalltown as the optimal location for the new contract manufacturer." State-to-state full truckload (FTL) truck rates are provided in Table 4. The truck capacities for different product components are provided in Table 5.

Just then, Abe McConnell entered the meeting. He was in charge of master planning for the production line. "Sorry I am late. I couldn't help but hear what you were saying, Julie. That is only partly true. The fact is that our product planners have a very difficult time planning the correct number of pieces to order from the overseas manufacturers." He went on to explain to Jim and Phil that the products DSM develops often either really take off or bust rather quickly. They frequently found that when a product took off, they lost potential sales because their initial inventory wasn't large enough as well as the fact that the supply chain was two to three months long before they could get any new product from China. On the other hand, their overall inventory write-off was greater than 10 percent because products they had thought would take off never did and their initial inventory was more than enough. In this case, they couldn't cancel the orders fast enough because there were too many months of inventory on the water on its way to Marshalltown. By now, Jim and Phil's heads were spinning. They couldn't dispute that Gustavo, Julie, and Abe were all making very good points. But how were they supposed to address them in their model? Particularly given the aggressive schedule and budget constraints imposed by the client? The first duty, though, was to build a baseline to establish understanding of the supply chain network.

Over the next six weeks, Phil and Jim communicated often with Julie, gathering data from both of the enterprise resource planning (ERP) systems used by the company—Oracle and SAP. Rectifying the calendar year's worth of data between all the files had been difficult, but they were finally ready to present a baseline network model to DSM. Additionally, some of the data requested, such as operating costs for the RDCs, capacity constraints of the RDCs, and labor costs for all facilities, were not able to be provided.

The purpose of the second meeting was to confirm that the baseline model was on the right track, and that the costs coming from the model were within 10 percent of DSM's estimated actual costs. This level of precision was necessary to ensure that any conclusions reached from their future modeling scenarios would be credible and stand up to scrutiny. The costs were indeed within their desired tolerances and the afternoon portion of that meeting consisted of one planned event and one surprise revelation.

Gustavo, Julie, Abe, and several others from DSM were all in a conference room brainstorming what various scenarios they wanted Jim and Phil to model. Jim was at the whiteboard writing down the various suggestions, which would later be prioritized. Some of the more plausible and popular scenarios included the following:

- Move the nearby Muscatine Contract Manufacturer with all the products it copacks to the West Coast, including Project Fragrance. The advantage of this would be to eliminate the need for imports to travel all the way to the Midwest and then for a percentage of the product to be sent back to the San Francisco RDC.
- Consolidate Shanghai and Hong Kong shipments at a single location for ocean transport.
- Rather than the current intermodal ground transportation network in the United States, determine the cost trade-offs between going all-truck versus all-rail.
- Combine all the activities of the various contract manufacturers into one facility.
- Allow the paint cans to be filled (i.e., produced) at the same location as the contract manufacturer, essentially reducing the transportation costs for the paint cans from a separate production point such as Marshalltown.

Each scenario was based on the assumption that DSM sourced a large brush from Asia, as well as a large paint can associated with it. The final scenario that DSM chose to model simply asked where the contract manufacturer (CM) should be located to minimize transportation costs should all CMs be combined into one. To evaluate these scenarios, many different candidate contract manufacturing and filling locations were proposed, and are illustrated in Exhibit 3. Exhibit 4 is an illustration of the number of demand points served by the RDCs given to the consultants by DSM.

As the brainstorming meeting was wrapping up, Julie approached Phil and Jim to say that Ken Martins, the global vice president for procurement and product development, had heard about the project they were working on and wanted to talk to them later that day.

The introductions between Ken and Phil and Jim had been cordial enough, but as soon as the meeting started, Ken started firing off many unanticipated questions. Had they thought of locating the contract manufacturer outside of the United States? What about Canada, near their Toronto RDC, or Mexico City, Mexico, near their Mexican RDC? When considering alternate filling locations for paint, had quality control and set-up costs been included in their costs? Were their analyses taking total product costs into account?

Ken's role had greater cost visibility compared with Gustavo, Julie, and Abe, who were focused primarily on minimizing costs for their respective components of the supply chain. This myopic focus had Ken wondering if other costs were higher than necessary because of this. He also wanted to know whether or not they could give him a definitive answer within the next three days, because there was yet another pending product, known by the code of "Compact Rain" for which he had to approve the funding by the end of the week. "Once again," he said, "I am faced with approving a contract manufacturer that is located within a 150-mile radius of Marshalltown for products that are being manufactured in Asia, to be sent first to Chicago, and then back to Marshalltown. It just doesn't feel right, but I can't get any data or network studies that show otherwise! I could use some new insight. So let me challenge you to be sure to consider all possible scenarios in your analysis."

Phil tried to explain that while Ken's questions were valid, they were outside the scope of the current project. Jim wondered, though, if perhaps part of the problem that DSM had was that the current study as well as previous studies had been scoped too narrowly to be able to really find the desired answers.

Before Jim and Phil flew back home, DSM treated them to dinner. During the course of the evening, a story was shared about the company's third-party logistics provider (3PL) company (a firm that helps manage the transportation aspect of moving products on behalf of DSM). The 3PL had recently sustained a major service disruption involving one of the vessels that DSM's goods were on. The West Coast port that the ship was supposed to have entered was experiencing a port strike and the ship would not be able to unload for several days. The 3PL had routed all the containers through Mexico instead and then trucked them through Mexico to the United States so that DSM could get the products with minimal delay.

Upon hearing the story, Julie commented, "Well, I hope that those extra costs didn't go against my budget. They should have been paid for by our Purchasing Department or the 3PL." It seemed later to Jim that perhaps Julie had misplaced incentives. She would rather the product sit for a couple of weeks in order to not affect her budget than have the product arrive on time. Could this also be an issue that DSM needed to deal with in finding the best way to design its network? Were misaligned incentives causing managers to lean toward the status quo?

Conclusion

All of these thoughts and memories were racing through Jim's mind as Phil eased the Mustang into DSM's parking lot for the final meeting. What would DSM think of the results they had? The results were certain to please some more than others. Jim wondered what other aspects could have been included in the model to help DSM.

Discussion Questions

1. Conduct an initial scenario cost analysis that will determine the optimal location for the new contract manufacturer. Take only into consideration the constraints and data that have been provided. Of the 26 locations that Jim and Phil tried, include at least 15 of them in your analysis. Des Moines, Iowa, should be included for comparison purposes (as being closest to Marshalltown). Additionally, include at least two of your own new possible locations.

    Please list your assumptions. Some assumptions will need to be made on your own to complete the analysis (such as distances between facilities). Examples of costs that were not provided and can be ignored are labor and warehousing costs.

2. Focusing on costs is the most obvious way to solve the problem. What are the various options and challenges independent of cost issues that are going on in this case? Be sure to consider supply chain trade-offs that might occur.

3. Ken issued a challenge to the consultants about considering all possible scenarios. What other scenarios could lead to a new optimal contract manufacturing location? (*Hint:* Consider relaxing a constraint to generate a new optimal location. Explain your choice and reasoning, and support with updated data analysis to determine if a new optimal CM location resulted.)

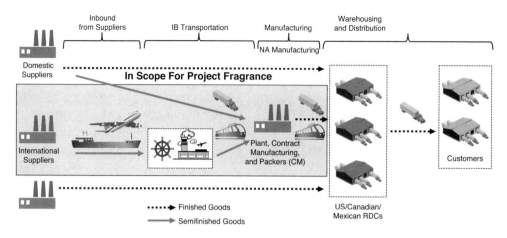

Exhibit 1 Scope of the project

Table 1 Ocean Lane Rates

Ocean Lanes	Miles	Cost per Mile / Container
Yantian—Long Beach	8,825	$0.34
Singapore—Vancouver	9,727	$0.34
Shanghai—Tacoma	6,964	$0.34
Shanghai—Seattle	6,951	$0.34
Shanghai—Oakland	7,480	$0.34
Shanghai—Long Beach	7,911	$0.34
Laem Chabang—Tacoma	9,135	$0.34
Laem Chabang—Savannah	11,362	$0.34
Laem Chabang—Long Beach	10,130	$0.34
Hong Kong—Tacoma	7,910	$0.34
Hong Kong—Long Beach	8,847	$0.34

The rates are standardized because DSM has a blanket contract that allows it to import to various ports with no cost difference. The only cost difference not included here would be specific port discharge fees.

Additional Data

All ocean containers are 40'. The volumes from Table 4 will need to be calculated in the appropriate ratio to determine how many ocean containers are sent annually. To convert a trailer to a 40' container (for ocean), a strict ratio of 40' to 53' may be taken for estimation purposes.

Volume: Brushes: 4,650,000 cases of brushes annually. Each case holds 15 brushes.

Volume: Paint cans: 34,875,000 cases of paint annually. Each case holds two cans.

Fuel surcharge per mode: Truck = 20% of gallon/gas cost. Rail = 2% of gallon/gas cost. If one gallon costs $5, then the fuel surcharge for trucks is $1.

Five-year growth volume predictions: California DC by 20%, Iowa DC by 15%, Georgia DC by 25%, Texas DC by 17%.

Table 2 Locations and Volumes of Contract Manufacturers and Supplier Locations by Port

Contract Manufacturing Locations	Contract Manufacturing Volume/Capacity	Supplier Locations	Supply Volume/Capacity
Muscatine, Iowa	1,000,000/1,000,000	Shanghai, China	1,250,000/1,600,000
Rochester, Minnesota	2,000,000/2,000,000	Hong Kong, China	2,575,000/3,750,000
Des Moines, Iowa	1,650,000/1,700,000	Bangkok, Thailand	825,000/825,000

All volume is in cases of products, not individual brushes or cans.

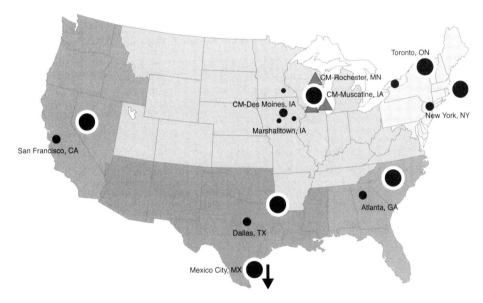

Example: Arizona customers are serviced by the RDC located in Dallas.
Diamonds denote contract manufacturer locations.

Exhibit 2 Service areas of each RDC (large circles)

Case 7 DSM Manufacturing: When Network Analysis Meets Business Reality

Table 3   Intermodal (Rail) State-to-State Costs, Excluding Fuel Surcharge

Origin State	Destination State	Transit Time (Days)	Rail Miles	Equipment Type	Line Haul Rate
CA	IA	5	2,080	53' Intermodal	$1,427
CA	NY	9	2,881	53' Intermodal	$2,792
CA	TX	7	2,101	53' Intermodal	$1,397
CA	GA	8	2,555	53' Intermodal	$2,585
WA	IA	6	1,890	53' Intermodal	$1,466
IA	CA	5	2,080	53' Intermodal	$1,577
IA	GA	5	803	53' Intermodal	$1,276
IA	NY	4	800	53' Intermodal	$1,414
IA	TX	3	1,041	53' Intermodal	$1,522

Additional lanes can be calculated by looking up mileages and taking the average cost per mile from the table above. (Fuel surcharge for rail = 2% of gallon/gas cost. If one gallon costs $5, then the fuel surcharge for rail is $0.10.)

Table 4   Full Truckload State-to-State Costs per Mile, Excluding Fuel Surcharge

To: / From:	CA	GA	IA	NY	TX	MN
AL	1.31	1.93	1.37	1.46	1.47	1.33
AR	1.32	1.35	1.32	1.31	1.62	1.31
AZ	1.13	1.1	1.13	1.18	1.2	1.1
CA	2.31	1.4	1.38	1.42	1.49	1.37
CO	0.98	1	0.93	1.08	1.06	0.91
CT	1.01	0.97	0.97	2.59	1.01	0.92
DC	1.4	1.38	1.37	1.94	1.44	1.33
DE	1.12	1.1	1.09	2.27	1.14	1.05
FL	0.93	0.91	0.89	0.96	0.93	0.85
GA	1.33	2.42	1.32	1.46	1.4	1.32
IA	1.63	1.72	3.22	1.84	1.81	2.21
ID	1.07	1.1	1.08	1.16	1.16	1.05
IL	1.42	1.5	2.01	1.67	1.51	1.95
IN	1.36	1.44	1.44	1.68	1.39	1.48
KS	1.39	1.41	1.47	1.46	1.57	1.36
KY	1.4	1.64	1.46	1.77	1.53	1.43

LA	1.42	1.49	1.43	1.56	1.88	1.4
MA	0.94	0.91	0.9	1.59	0.95	0.87
MD	1.13	1.13	1.1	1.8	1.18	1.09
ME	1.44	1.44	1.39	1.96	1.47	1.35
MI	1.31	1.34	1.3	1.57	1.34	1.36
MN	1.58	1.61	2.2	1.74	1.69	2.24
MO	1.38	1.43	1.67	1.45	1.51	1.4
MS	1.27	1.49	1.33	1.41	1.54	1.3
MT	0.86	0.97	0.93	1.04	0.99	0.87
NC	1.3	1.42	1.21	1.54	1.21	1.21
ND	1.21	1.22	1.27	1.32	1.29	1.17
NE	1.57	1.59	2.09	1.73	1.73	1.59
NH	1.02	0.99	0.97	1.55	1.02	0.93
NJ	1	0.99	0.96	2.58	1.02	0.94
NM	1.03	1.06	1.02	1.15	1.13	1.04
NV	1.56	1.44	1.41	1.5	1.52	1.38
NY	1.17	1.17	1.11	2.65	1.21	1.09
OH	1.45	1.54	1.48	2.02	1.53	1.49
OK	1.41	1.44	1.43	1.52	1.86	1.4
OR	1.14	1.1	1.06	1.15	1.14	1.03
PA	1.12	1.13	1.11	2.15	1.15	1.08
RI	0.8	0.79	0.77	1.31	0.82	0.74
SC	1.24	1.89	1.26	1.39	1.31	1.24
SD	1.28	1.3	1.42	1.43	1.37	1.27
TN	1.18	1.62	1.24	1.43	1.33	1.22
TX	1.13	1.16	1.11	1.22	1.67	1.1
UT	1.01	1	0.97	1.08	1.09	0.97
VA	1.23	1.29	1.25	1.64	1.31	1.24
VT	0.98	0.94	0.92	1.65	0.98	0.87
WA	1.05	1.04	1.01	1.08	1.06	0.98
WI	1.63	1.71	2.32	1.89	1.74	3.76
WV	1.5	1.62	1.52	2.1	1.6	1.47
WY	1.09	1.15	1.11	1.25	1.22	1.09

(Fuel surcharge for trucking = 20% of gallon/gas cost. If one gallon costs $5, then the fuel surcharge for trucks is $1.)

Table 5 Weights and Volumes of Product on the Trailers

53' Trailer	Cases/ Pallet	Cases/Trailer (Cases per Pallet x Pallets)	Total Product Weight (lbs)	Pallets/ Trailer	Total Pallet Weight (lbs)	Total Weight/ Trailer (Product Wt + Pallet Wt)
Brush	150	6,600	19,800	44	2,948	22,748
Paint can	90	1,980	39,600	22	1,474	41,074
Package*	90	1,980	40,392	22	1,474	41,866

* A package is one can and one brush packaged together.

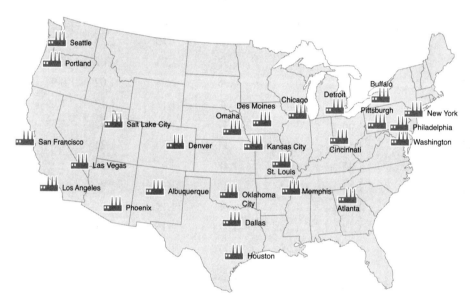

Exhibit 3 Alternate CM locations tried in the model by Jim and Phil

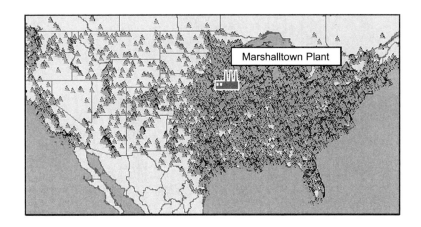

Exhibit 4   Demand points serviced by the RDCs for visual reference only

8

KIWI MEDICAL DEVICES, LTD.: IS "RIGHT SHORING" THE RIGHT RESPONSE?

Stanley Fawcett, Weber State University

The Race of Life

Exhausted, Michelle Ledger leaned against the Stop sign, stretching out her burning leg muscles. She had just finished a 90-minute, 20-kilometer run. She wasn't training for a marathon; rather, she was burning off anxiety. Every time pressure at work ratcheted up, she turned to her favorite pastime, long-distance running. Somehow, running along Waitemata Harbour and looking to the Waitekere Range calmed her inner worries as she focused on the task du jour at work. As she began her cooldown routine, Michelle reflected on the events of the past few months.

At the forefront of her thoughts was the morning's executive steering committee meeting. It had not gone well. Timothy Craig, CEO of Kiwi Medical Devices, Ltd., had pressed Michelle for a solution to Kiwi's eroding market share. Michelle hated having to say, "I don't know," but she didn't have an answer. Worse, she had to admit she wasn't sure when she *would* know. Kiwi had never offshored and those decisions were complex and interrelated. Yet, despite the angst caused by Tim's repeated, intense questioning, Michelle was certain that a wrong decision made in haste would cost far more than the uneasiness of the moment.

As director of operations at Kiwi, Michelle had been tasked with increasing Kiwi's manufacturing capacity and competitive capabilities. The need to evaluate offshoring was raised by a recent market analysis, which revealed that although Kiwi retained an innovation edge over aggressive global rivals, it had lost cost and delivery advantages. As a result, competitors were gaining shares in markets around the world. By setting up production operations in Asia, competitors had lowered their cost structures, changing the rules of the competitive race. Michelle and her team needed to respond—and quickly. Although confident her team could help Kiwi regain its leadership position, Michelle wasn't quite sure how Kiwi could best restructure its global operations to gain advantage.

Kiwi's Marathon Begins

Kiwi Medical was born in the late 1960s. The offspring of Kiwi Electronics—an innovative maker of household appliances—Kiwi Medical came into existence as a hedge against a highly volatile appliance market. Kiwi Electronics had sought to find a counter-cyclical market where it could leverage its technological expertise. Heated humidification devices used in respiratory and sleep apnea applications seemed to offer a good fit from at least three perspectives.

- The industry was underdeveloped and lacked established competitors.
- Kiwi's technology and research and development (R&D) expertise gave it a strong foundation that could be used to transform the industry and establish Kiwi as a world leader.
- The medical device industry had strong global growth potential.

Following 30 years of solid growth and good financial performance, Kiwi Medical was spun off as an independent company in 2001. Kiwi's headquarters and manufacturing were in Auckland, New Zealand.

Kiwi's Sprint for Global Sales

By 2009, Kiwi Medical's global sales had reached NZ$485 (its market capitalization was approximately NZ$1.5 billion). Kiwi sold to hospitals, long-term care facilities, and home health-care dealers in more than 120 countries. Its core products included respiratory humidifiers and neonatal care products, including infant warmers and resuscitators. Kiwi also enjoyed a strong presence in the obstructive sleep apnea (OSA) market with a focus on continuous positive airway pressure (CPAP) devices. Kiwi also manufactured and distributed the accessories needed to deploy its equipment (e.g., single-use and reusable chambers as well as breathing circuits). See Exhibit 1 for a breakdown of sales by product group.

From 2005 to 2009, sales grew at a solid, if not spectacular, pace of 19 percent per year (see Table 1). Kiwi attributed the sales growth to its intense focus on technological innovation and a relentless quest to expand its global market presence:

- **Product innovation**—Kiwi was dedicated to improving existing products and developing innovative, complementary products. Indeed, a continued commitment to research and development enabled Kiwi to target new medical applications for its core technologies. From 2005 to 2009, Kiwi spent an average 6.3 percent of sales on product development and clinical research. By 2009, its research team of engineers, scientists, and physiologists had grown to 273 people. In 2009 alone, Kiwi had obtained 393 patents in markets around the world (82 in the United States). An additional 369 patent applications had been filed (77 in the United States).

- **Global market development**—Kiwi's largest market was North America; however, Kiwi had a strong presence in both Asia and Europe (see Exhibit 2). To support its aggressive marketing, Kiwi developed a 500-person sales, marketing, and distribution team. Direct sales offices had been established in Australia, China, the Euro zone, India, New Zealand, Scandinavia, the UK, and the United States. Kiwi had also built relationships with more than 90 distributors worldwide. To support continued market expansion, Kiwi opened two new distribution centers in Canada and Japan in 2009.

Kiwi's Cramped Manufacturing Footprint

Even as Kiwi chased global sales, operating costs seemed to be racing out of control—they had increased 16 percent per year over the past five years. Kiwi had long been proud of its New Zealand heritage. Kiwi touted, "We manufacture, assemble, and test our complete range of products, including many components, in our custom-built facilities in New Zealand with a total area of approximately 51,000 square meters." These facilities possessed advanced manufacturing technologies and had obtained highly visible ISO9001 and ISO13485 (international medical device) quality certifications.

Although provincial, New Zealand production had provided Kiwi a relative cost advantage vis-à-vis its toughest rivals in Europe, Japan, and the United States. However, now that rivals had globalized their manufacturing networks, Kiwi's limited base of operations was becoming a liability.

Kiwi's Race Turns Uphill

By most measures, 2009 had been an outstanding year for Kiwi. Operating revenue was up 28 percent with both product groups—respiratory (32 percent) and OSA (25 percent)—delivering strong growth. Better yet, operating profit increased by a very healthy 76 percent. However, a very challenging 2008 coupled with the general upward trend in operating costs led Tim Craig to initiate an in-depth analysis of competitive and market trends.

Michelle had been a member of this scanning team. To her dismay, many of the findings—especially those related to Kiwi's more mature respiratory market products—had led her back to her favorite running trail along Waitemata Harbour. She kept a list of the most pertinent findings posted by her office door. She had highlighted the most distressing points in red. The list included the following items that would influence Kiwi's long-term success in the respiratory market:

- The respiratory device market was a NZ$1.4 billion market. Global demand for respiratory care devices was expected to rise 3–5 percent annually for the next 20 years.

- Kiwi's share of the respiratory market had decreased from almost 25 percent to 17.5 percent in the past four years. Kiwi's market share was falling by 2 percent per year!

- Kiwi's rivals were headquartered in Europe, Japan, and North America. Although their historical costs had exceeded Kiwi's, four of Kiwi's five largest competitors had shifted production to low-cost sites in China, Vietnam, and Indonesia. As a result, these rivals now enjoyed a 10–15 percent price advantage for comparable respiratory devices.

- New low-cost, but less technologically sophisticated rivals had emerged in China. These rivals could not match Kiwi's innovation, but they put tremendous pressure on Kiwi's more mature products. Of note, these Chinese rivals challenged Kiwi's value proposition. Some customers had begun to ask a threatening question, "For some applications, how much price premium is state-of-the-art technology worth?"

- Real wages in New Zealand were rising faster than in most of Europe and North America. Collective bargaining units (unions) were gaining a foothold with Kiwi's former parent company, Kiwi Electronics.

- Kiwi Breathers (the name of Kiwi's respiratory devices) were crated and shipped airfreight in lots of five on an ex-works (EXW) basis to customers outside New Zealand. Although product could be delivered to 80 percent of Kiwi's global customers within five to seven days, customers were asking for even faster delivery. Many customers were now ordering competing devices from more geographically proximate rivals.

- Available airfreight capacity from New Zealand was dropping. Transportation costs from New Zealand were rising faster than from North America or Europe. *(Were freight rates during the 2007 oil spike a harbinger?)*

The competitive marketplace was clearly more hostile and volatile than at any time in Michelle's nine years with Kiwi. After taking a deep dive into the facts presented by the market study, Kiwi's management determined that it needed to reduce operating expenses by at least 3 percent per year over the next five years while offering better (faster, more consistent) delivery. The goal was to reverse Kiwi's market share slide and raise operating profit by 5–8 percent annually. Offshoring seemed to be the only possible response. As Michelle and her team began to explore this option, they identified three separate, but related questions that needed to be answered:

1. Where was the right place to set up overseas manufacturing?
2. How much investment and risk was Kiwi willing to incur as it set up overseas operations?
3. How should Kiwi provide logistical support for its overseas operations?

Choosing Where to Run

For Michelle and her team, which had become known as the "right-sourcing" team, the natural starting point in the search for a right-sourcing location was to follow the competition to Asia. However, given Kiwi's global sales platform and customers' desire for responsive delivery, the team decided to add Slovakia, Poland, and Mexico to the initial set of possible host countries. As the right-sourcing team met to discuss options, they began to identify potential country selection criteria. From an initial list of more than 30 criteria, the team settled on the following 10 criteria that appeared to best define Kiwi's needs:

1. Labor costs
2. Labor skill and experience
3. Transportation costs
4. Transportation lead times to/from the factory
5. Duty rates for incoming raw materials as well as export of finished product
6. Political stability/corruption
7. Local taxes

8. Unions, strike risk
9. Permits and factory set-up cost
10. Management relocation expense, lifestyle, and safety

As the right-sourcing team defined the critical criteria, several countries under consideration fell off the list, leaving the team with four finalists to evaluate more closely: China, Indonesia, Mexico, and Slovakia. To help make the final decision, the team assembled a fact sheet on each country (see Table 2).

Despite the methodical approach, different members of the team adopted advocacy roles for each of the countries. Some team members felt it would be foolish to allow competitors to operate in China unimpeded. They did not want to give rivals a first-mover's advantage in developing the Chinese market. Others noted that Kiwi's strength was leading-edge innovation. They felt being close to developed markets of Europe and the United States was key. Michelle could understand both arguments and hoped that rigorous analysis would provide a tipping point that all of the team could rally around.

Once a country was selected, Michelle knew it would be vital to select the right city. This decision, however, was more important and challenging for China and Mexico. After all, the options in these two countries were greater and more diverse. In both countries, the team noted that the most important difference was between coastal/border investment zones and interior cities. The advantage of investment centers like China's Guangdong province and Mexico's border cities of Tijuana and Juarez was simplified setup and logistics. By contrast, interior cities like Chongching and Saltillo offered lower wage rates, a more-abundant and stable workforce, and better tax incentives.

Running with Risk, but How Much?

From the inception of the analysis, Michelle had recognized that neither she nor anyone else on the right-sourcing team had global manufacturing experience. Even so, Michelle was not so naïve to think business the New Zealand way could be easily transferred to one of the finalist countries. She realized that regardless of country choice, navigating the politics and managing the cultural differences in the country of choice would be a real challenge. This reality increased both Kiwi's and Michelle's risk of failure. No wonder she spent so much time running during the past three months!

Fortunately, as the team invested in the due-diligence process, they identified opportunities to minimize Kiwi's exposure to the uncertainties of global operations. Most important, they realized that going global did not mean they had to go all in. Three options for expanding capacity and improving competitiveness existed: subcontracting, shelter operations, and a wholly owned subsidiary.

Subcontracting offered the easiest and the fastest way (often as little as 30 days) to get started. Subcontracting involved finding a local manufacturer of original equipment manufacturer (OEM) electronics that could produce to Kiwi's design specifications. Kiwi could thus minimize investment. Although Kiwi would have to provide the specialized equipment and arrange for delivery of needed components, it would have no brick-and-mortar responsibilities. The subcontractor would manage manufacturing and provide logistics support. Two downsides worried Michelle. Kiwi would lose some control over the production process and product reliability. Further, she expected that unit costs would be higher under subcontracting.

The *shelter option* required more time to get going than subcontracting (around 60–90 days), but compared with a wholly owned subsidiary, shelters reduced the initial investment and mitigated the learning challenges associated with new setups. A "shelter" service provider would manage the hassles of getting the business up and running. In essence, Kiwi could maintain control of the production process and product technologies without getting bogged down in administrative and legal annoyances. As the team met with several shelter service providers, five advantages of shelter operations caught their attention:

1. Assistance with accounting and tax services as well as obtaining licenses and permits
2. Help with legal and fiscal representation in the host country
3. Hiring workers and providing HR services, including payroll and performance monitoring
4. Help sourcing raw materials and managing third-party warehousing
5. Help with customs clearance and duty-rate analysis

Wholly owned subsidiaries offered Kiwi maximum control and they often delivered the lowest operating cost structure, but they were know-how intensive and risky. If Kiwi pursued this option, the right-sourcing team would have to find a site, manage construction of the facility, and perform all of the administrative tasks performed by the shelter service providers. Understanding the details of doing business locally was a must. So too was developing needed political and business relationships. Ownership was the preferred option for companies (1) looking to make a long-term commitment, (2) seeking to establish large-scale operations, and (3) needing high levels of technological or new-product support. The key was achieving the long-term success needed to justify the up-front capital and emotional investments.

Selecting a Race Support Team

Even as the team began to develop the country-selection criteria, Michelle realized that right-sourcing demanded more than just good manufacturing decisions. The fact that several of the criteria were transport-oriented underscored the need to build the right logistics infrastructure to support the overseas operation. The good news: Kiwi had a lot of exporting experience. At least here, the team would not be running completely in the dark. Still, supporting a global manufacturing operation would be more complex than anything Kiwi had done before.

Because of time pressures and the lack of direct experience, the team decided the best option would be to outsource logistics support to a third-party logistics (3PL) company. Such a decision fit well with Kiwi's past behavior. Senior management at Kiwi had always emphasized that Kiwi was an R&D and manufacturing company—not a logistics provider.

To prepare for effective outsourcing, the team mapped out the basic materials flows. Mapping revealed the need to find 3PLs that could handle three distinct types of logistics: (1) inbound movement of capital equipment, (2) inbound movement of raw materials and components, and (3) outbound movement of finished goods to customers around the world. A closer inspection of the materials flow suggested that Kiwi should evaluate the 3PLs' ability to set up and manage an inbound cross-dock warehouse. After all, regardless of country choice, Kiwi's primary suppliers were geographically dispersed and would be shipping product into a port of entry—probably in less-than-container-load quantities. From a cost perspective, it might make sense to consolidate these shipments at the entry point and then ship them on a truckload basis to the new factory.

Based on this analysis, the right-sourcing team developed a request for information (RFI) for the following:

- Pricing, capacity, and lead-time information for movement of raw materials from 10 international suppliers to the appropriate port of entry for each of the four country finalists
- Pricing and timing information for movement of capital equipment from five European suppliers to each proposed site option
- Pricing for consolidation of inbound shipments at a cross-dock warehouse
- Pricing, capacity, and lead-time information for appropriate ground transport from port of entry to actual manufacturing site
- Pricing, capacity, and lead-time information for appropriate ground transport from factory to airport for export shipping

- Pricing, capacity, and lead-time information for movement of finished product to international customers by either truck or air depending on customer location
- Pricing information for preparing all paperwork (customs clearance as well as inspection of capital equipment and raw materials) on inbound shipments to the factory
- Pricing information for all import/export documentation related to outbound shipments of finished goods (to include documents for international shipments)

The team had just sent out the RFIs a few days earlier. The three big logistics integrators—DHL, FedEx, and UPS—had been selected as potential support team leaders because they were viewed as capable of providing one-stop shopping for Kiwi's global needs. Although DHL had been Kiwi's sole global 3PL for its New Zealand import/export shipments for 20+ years, the team felt it was time to verify that DHL was still capable of providing Kiwi top-notch service at the lowest possible prices.

In addition, tailored RFIs were sent to local freight forwarders in each country to assess their ability to provide ground transport, cross-docking, and customs clearance/documentation services. Kiwi wanted to make sure that it built the right relationships to get things done on the ground. Sometimes this could be done best by a local player with "good" connections.

Time to Relax—For a Moment

Having run through the day's stress, Michelle finished her cooldown routine. Suddenly, she realized that until the detailed responses to the RFIs were returned, the team could do little more than wait. The rest of the homework had been done. The right-sourcing team was closer to making more progress than she had realized. Life was good after all.

Of course, Michelle knew that this was the lull before the storm. Once the information from the 3PLs came back, a decision would need to be made. Then the real work of execution would begin. Only then would they find out whether they had truly right-sourced Kiwi's mature products, reducing freight and manufacturing costs, shortening lead times, and increasing Kiwi's geographic reach. If they had done their job well, Kiwi would have more capacity at its New Zealand facility to design and build the new, more-technologically intensive products that were the lifeblood and future of Kiwi Medical Devices, Ltd.

Discussion Questions

1. Which of the four finalist countries should Michelle and the right-sourcing team select for its new right-shore facility? Or should Kiwi stop, take a deep breath, and continue its New Zealand-centric manufacturing model? *Hint:* A weighted-factor model might provide insight into a good choice.

2. Given Kiwi's competitive challenge and its strategic goals, would you suggest a port of entry or an interior city for the new manufacturing facility? What specific factors drove your decision?

3. Which mode of entry—subcontracting, shelter, or wholly owned subsidiary—do you endorse? What factors should drive this decision?

4. What do you think of Kiwi's decision to invite new players to bid for the logistics support business associated with the right-shore operation?

5. Role-play opportunity: Consider yourself the newest member of the right-sourcing team. Focusing on the information presented in Table 2, use a spreadsheet to develop a weighted factor model to help decide among the country options. Use the narrative from the case to help establish appropriate weights for each criterion. Perform a sensitivity analysis and come to class prepared to share your findings. Beyond identifying the "right" country for Kiwi's new manufacturing facility, be sure you can answer one of Timothy Craig's favorite questions: "How robust is your solution?"

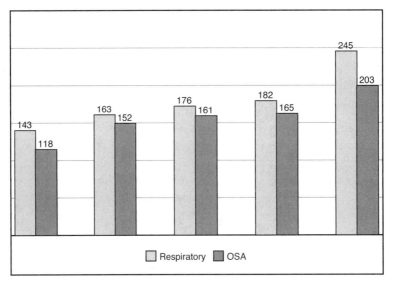

Exhibit 1   Sales by product group (NZ$000)

Table 1 Financial Performance at Kiwi Medical Devices, Ltd.

	2009	2008	2007	2006	2005
Sales Revenue (NZ$000)	538,923	384,022	380,706	321,397	267,028
Foreign exchange gain (loss)	−29,747	13,239	4,639	38,750	34,237
Total Operating Revenue	509,176	397,261	385,345	360,147	301,310
Cost of goods sold	235,417	197,370	167,941	134,715	108,921
Sales, general, & other expenses	132,011	108,623	106,459	90,664	74,794
Research & development expenses	31,424	26,741	22,941	19,256	17,978
Operating Profit	113,654	64,527	88,004	115,512	99,617
Net financing expense	−19,262	−4,242	374	389	1,384
Profit before tax	94,392	60,285	88,378	115,901	101,634
Tax expense	−25,314	−21,128	−32,228	−38,240	−33,474
Profit after tax	69,079	39,157	56,150	77,661	68,160
Revenue by Product Group:					
Respiratory & acute care	271,425	202,068	195,305	180,416	158,663
Obstructive sleep apnea	224,890	183,570	178,775	168,887	131,285
Distributed and other	12,860	11,624	11,264	10,845	11,361
Total	509,175	397,262	385,344	360,148	301,309
Revenue by Region:					
North America	231,836	183,910	188,355	187,157	146,798
Europe	168,617	128,759	119,811	95,927	86,488
Asia Pacific	79,684	64,699	59,287	60,668	55,968
Other	29,040	19,893	17,891	16,395	12,056
Total	509,177	397,261	385,344	360,147	301,310
Financial Position:					
Tangible assets	412,387	348,170	338,864	281,825	236,797
Intangible assets	46,861	19,980	23,856	15,067	18,847
Total assets	459,248	368,150	362,720	296,892	255,644
Liabilities	−232,474	−150,963	−100,153	−70,459	−42,490
Shareholders' equity	226,774	217,187	262,567	226,433	213,154

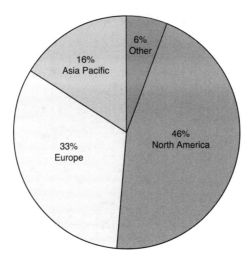

Exhibit 2   Kiwi's sales by region (2009)

Table 2   Country Fact Sheet

China is the world's fourth largest country with 9,596,961 square kilometers in area and the most populous country with 1.3 billion people (median age = 34.1; fertility rate =1.79). China has a rich heritage and history—its name translates to middle kingdom. Since the early 1990s, China has initiated economic reforms that have led to dramatic industrialization. Average gross domestic product (GDP) growth has exceeded 10% for the past 20 years, helping lift GDP/capita to NZ$9,168. Since 2000, no country has attracted more foreign direct investment (FDI) than China.

Indonesia is a Southeast Asian archipelago consisting of hundreds of islands that comprise 1,904,569 square kilometers. Indonesia is the world's fourth most populous and largest Islamic country with 240 million people (median age=27.6; fertility rate=2.31). GDP/capita (PPP) is NZ$5,641 and has recently grown at more than 5% per year. About 17.8% of the population lives below the poverty line. As a member of ASEAN and one of Asia's emerging Tigers, Indonesia is an attractive FDI destination; yet, an inadequate infrastructure hinders economic growth.

Mexico has a landmass of 1,964,375 square kilometers, and a population of 111 million (median age=26.3; fertility rate=2.34). Mexico shares a 3,200-km border with the United States. Since ratification of NAFTA in 1994, Mexico has achieved an impressive record of attracting FDI, becoming one of the most industrialized countries in Latin America. GDP/capita is NZ$18,618. The labor force is estimated to be 46 million, unemployment is at 6.2%, and 18% of the population lives in poverty. Violence has increased since President Calderon declared war on Mexico's drug cartels.

Slovakia is a small country (49,035 square kilometers) located in central Europe with a population of 5,463,046 (median age=36.9; fertility rate=1.35) As a member of the EU, Slovakia has ready access to the rest of the European market. Economic reform has led to rapid industrialization and growth. GDP/capita is NZ$29,760. The labor force is 2.6 million with an unemployment rate of 11.8% and 21% of the population living below the poverty line. The literacy rate is 99.6%.

Criteria	China	Indonesia	Mexico	Slovakia
Hourly Compensation	NZ$1.14	NZ$1.07	NZ$5.51	NZ$11.79
Labor Skill	Good	Acceptable	Very Good	Very Good
Transportation Costs:				
New Zealand to…	Low	Low	Medium	High
…to Asia	Low	Low	Low	High
…to Europe	High	High	Medium	Low
…to U.S.	Medium	High	Low	Medium
Transport Lead Times:				
New Zealand to…	Very Good	Excellent	Good	Good
…to Asia	Excellent	Excellent	Good	Good
…to Europe	Acceptable	Acceptable	Good	Excellent
…to U.S.	Acceptable	Acceptable	Excellent	Good
Duty Rates:				
New Zealand to…	Low, Variable	Low, Variable	Free w/ re-export	Low, Variable
…to Asia	No duty	No duty	3–4%	3–4%
…to Europe	Free/19.6% VAT	Free/19.6% VAT	Free/19.6% VAT	Free/19.6% VAT
…to U.S.	No duty	No duty	No duty	No duty
Corruption Index	3.6	2.8	3.3	4.5
Economic Freedom Index	51	55.5	68.3	69.7
Global Competitiveness	4.74	4.26	4.19	4.31
Taxes (Corporate/)	25%/17%	25%/10%	28%/16%	19%/19%
Strike Risk	Low, increasing	Moderate	Moderate	Moderate
Factory Setup:				
Permit (ease/cost factor)	Moderate/High	Easy/Moderate	Easy/Moderate	Moderate/High
Land Cost	NZ$42/m^2	NZ$42/m^2	NZ$40/m^2	NZ$54/m^2
Building Cost	NZ$279/m^2	NZ$317/m^2	NZ$247/m^2	NZ$396/m^2
Management Lifestyle	Acceptable	Difficult	Favorable	Favorable

Part 3

RISK AND UNCERTAINTY IN THE SUPPLY CHAIN

This section contains two cases: (1) "Innovative Distribution Company: A Total Cost Approach to Understanding Supply Chain Risk" and (2) "Humanitarian Logistics: Getting Donated Foods from Switzerland to Zambia." Although the focus of the two cases is best described as addressing risk and uncertainty, each case examines a particular aspect of risk and uncertainty, which particularly challenges managers.

The "Innovative Distribution Company: A Total Cost Approach to Understanding Supply Chain Risk" case focuses on the many activities in supply chain management (SCM) that may be quantified to assist with a lowest total cost of ownership decision that explicitly considers supply chain risk. The case applies the analysis in two different supply chain scenarios: one global and one domestic. The case requires calculation of economic order quantity (EOQ) and safety stock quantities, and then the combination of purchase price, shipping costs, and inventory carrying costs to quantify the differences between the two supply chains. The case illustrates the impact of differences in labor rates (and thus outsourcing), use of exchange rates, INCOTERMS versus free on board (FOB) terms, and a variety of risk factors, such as piracy, capacity, natural disasters, damage in transit, port congestion, obsolescence, and so on.

The "Humanitarian Logistics: Getting Donated Foods from Switzerland to Zambia" case focuses on the increasingly important global topic of relief logistics and its attendant risks and uncertainty. As natural disasters, wars, food shortages, and growing unemployment are more easily communicated throughout the world, the awareness of the challenge of humanitarian logistics gains greater visibility. The case illustrates that although humanitarian logistics benefits from the knowledge and insights developed in commercial and military logistics, many unique aspects do and will continue to exist: little advance warning, no established distribution channels, harsh climates, remote areas with little to no infrastructure, politics, the threat of disease and violence, and so on. The case requires the investigation of shipment mode and cost, routing, documentation, use of INCOTERMS, exchange rate fluctuation, ownership of transportation, customs, and other typical challenges of global business, which are exacerbated by the nature of humanitarian logistics.

9

INNOVATIVE DISTRIBUTION COMPANY: A TOTAL COST APPROACH TO UNDERSTANDING SUPPLY CHAIN RISK

Dr. Ted Farris, University of North Texas
Dr. Ila Manuj, University of North Texas

Introduction

This case illustrates the use of the total cost of ownership concept to analyze two supply chains, one international and one domestic. Students must calculate economic order quantity and safety stock quantities, then combine purchase price, shipping costs, and inventory carrying costs to quantify the differences between the two supply chains.

The domestic versus international aspects of the case allow the instructor latitude in discussing:

1. Differences in labor rates driving outsourcing
2. Use of exchange rates
3. Contrast INCOTERMS versus free on board (FOB) terms
4. Understand the economic development of inland China and its developing infrastructure
5. In-transit carrying costs and onsite inventory carrying cost

6. Economic order quantity
7. Safety stock
8. Total cost of ownership

Learning Objective and Appropriate Audience

The case addresses many activities in supply chain management that may be quantified to help assist a lowest total cost of ownership decision. It has been effectively used with the intermediate to advanced student in a senior-level capstone course to synthesize the many trade-offs that should be considered in supply chain management. It may also be effectively utilized within a junior-level international logistics course.

"Arrrgh!" exclaimed James L. Heskett, president of Innovative Distribution Company (IDC). "Pirates have struck again off the coast of Somalia. It seems like every time we turn around, there is another piracy on the High Seas."

"Unfortunately, that is nothing new," replied John L. Hazard, VP of Supply Chain Excellence. "Piracy has been going on for centuries and is still going on today. Did you know piracy has been dramatically increasing? In 2005, there were 276 piracy incidents[1] and in 2009 there were 406 incidents worldwide."[2]

"Wow! That has got to cost someone a bundle. Who pays for that?" asked Heskett.

"I read a segment on MSN about that,"[3] responded Hazard, "and the cost of insuring ships has gone up. Insurance premiums increased by 10 times in 2009. Some companies are spending more time training their crews, whereas others are avoiding the area altogether—taking long trips around Africa's southern tip can add 2,700 miles to each trip and increase fuel costs by $3.5 million annually. And, because the ships can only make five round trips per year instead of six, delivery capacity has dropped by 26 percent. Who pays? The customer!"

"Gee, I never thought of those costs. The supply chain really takes a hit. It is a good thing we do not ship anywhere around Somalia," exclaimed Heskett.

"But there is risk everywhere," challenged Hazard. "Piracy occurs around the world. They have piracy problems in Malaysia and off the coast of Brazil as well. And there are lots of other risks in the supply chain that need to be mitigated. We have embraced off-sourcing because of lower unit prices, but we need to consider the total cost of ownership of the supply chain. Longer transit times, fluctuating exchange rates, uncertain delivery schedules, disruptive weather patterns, multilanguage requirements, political turmoil, unique tariffs, and duties all add to the cost of doing business internationally. I'm not sure we understand the true cost of our supply chain."

"You have a great point. We ought to take a look at all the costs of sourcing IDC's next new product—Schachtel Schmuggel Bannware—and consider the entire supply chain costs," pondered Heskett. "See what numbers you can gather and we'll take an all-in look at the numbers."

A few days later, Hazard and Heskett met to review all of the information they had gathered about the new product.

New Product Sourcing Details

"What did you find?" asked Heskett.

"There are only two possible sources of supply for IDC's new product. We cannot buy or hold fractional units of a product and we have a projected annual demand (based on a 365-day year) of 21,500 units with a deviation in daily sales of 11 units. Our goal is to maintain an in-stock probability of 97.7 percent for our customers," replied Hazard.

"All product (regardless of supplier) will be shipped by rail, utilizing 20-foot equivalent units (TEUs) to IDC's distribution center in Alliance Fort Worth (AFW) where we will service all of IDC's customers' needs. A single TEU container can hold up to 600 units of Schachtel Schmuggel Bannware. Due to the nature of the product, no other product may be loaded into the same container. IDC's inventory carrying cost throughout the supply chain is 32.2 percent."

Hazard and Heskett recognize it will cost $105 to place each order with the domestic supplier and due to the complexity of international trade will cost $182 to place each order with the foreign supplier.

Domestic Supplier Details

One of the possible sources of supply is CousinsAg, located in Wahoo, Nebraska. The U.S. Department of Labor reported that in 2002, 88,000 of Nebraska's wage and salary workers were members of unions.[4] CousinsAg is a union shop with an average labor rate in their Wahoo, Nebraska, facility of $25.30 per hour. In responding to IDC's Request for Quote (RFQ), CousinsAg's price is $85.00 per unit.

As shown in Exhibit 1, when an order is placed with CousinsAg, it will take 10 days for them to process and manufacture the order, and an additional 5 days to ship it FOB Origin Prepaid to IDC's Alliance Fort Worth Distribution Center. Rail shipping cost from CousinsAg to AFW is $1,850. Based on similar rail shipments from that part of the country, Hazard assumes the standard deviation of the shipping time from Wahoo will be 1.14 days.

Global Supplier Details

The other possible source of supply is Dong Hai Supply, in Chengdu, Sichuan, China. Over the past decade, China aggressively developed its transportation and logistics infrastructure inland from the coast. As shown in Exhibit 2, the Chinese government is now actively promoting trade in areas such as Chengdu. Located 2,107 kilometers from the port of Shanghai, the Sichuan Administration of Price Control, Sichuan Department of Finance, and the Sichuan Labor Department have maintained strict wage controls to help develop manufacturing for export. The average labor rate in Chengdu is 10.36 Yuan per hour. The current exchange rate is 1 CNY China Yuan Renminbi (¥) = 0.14646 U.S. Dollar.[5] In responding to IDC's Request for Quote (RFQ), Dong Hai Supply's price is 547 ¥ per unit.

The global supply chain is shown in Exhibit 2. When an order is placed with Dong Hai Supply (EXW Chengdu, China), it will take 15 days for them to process, manufacture, and stuff the order into a TEU container. Dong Hai Supply will use the Interface Exporting Company (IEC) to ship the container FCA Long Beach.

As a part of China's aggressive development in infrastructure, the high-speed Shanghai-Chengdu Railroad has recently been completed,[6] and will take IEC one day to move the container by rail from Chengdu to Shanghai. It will wait four days at the Port of Shanghai waiting to be loaded onto a ship, 16 days to cross the Pacific Ocean to the Port of Long Beach, and three days waiting to clear customs and be unloaded onto a dockside rail spur in Long Beach.

IEC charges ¥ 12,414.5 for each TEU shipped. Import tariffs and duties are $325 per TEU and are incurred at Long Beach U.S. Customs and charged separately to IDC on a monthly basis. Once the shipment clears customs and is off-loaded to railcar in Long Beach, it will take an additional four days to ship it FOB Origin Prepaid to IDC's Alliance Fort Worth Distribution Center. Rail shipping cost from Long Beach to AFW is $2,250. Based on similar mini-landbridge shipments from inland China, Hazard assumes the standard deviation of the shipping time will be 3.45 days.

Faced with this information, Heskett has asked Hazard the following questions.

Discussion Questions

1. Using the current exchange rate, what is the *initial purchase cost per unit* (in U.S. dollars) paid to Dong Hai Supply? (Do not include transportation costs.)

2. What is the *average time* for an order filling a TEU container to come from Dong Hai Supply in Chengdu, China, to IDC's Alliance Fort Worth Distribution Center? From CousinsAg in Wahoo, Nebraska, to IDC's Alliance Fort Worth Distribution Center?

3. Using the current exchange rate, what is the *cost* (in U.S. dollars) to ship a TEU container from Dong Hai Supply in Chengdu, China, to IDC's Alliance Fort Worth Distribution Center?

4. What is the *economic order quantity* (use unit price only; do not include transportation costs) if we purchase everything from CousinsAg? From Dong Hai Supply?

5. How many units of *safety stock* will we need to hold if we purchase everything from Dong Hai Supply? From CousinsAg?

6. Inventory carrying costs are based on the value of the product at the time it is held in inventory. What is the *in-transit carrying cost per unit* (in dollars and cents) if we purchase everything from Dong Hai Supply? From CousinsAg?

7. What *average inventory level* (in units) will we hold at the IDC's Alliance Fort Worth Distribution Center if we purchase everything from Dong Hai Supply? (Be sure to consider both safety stock and cycle stock.) From CousinsAg?

8. Inventory carrying costs are based on the value of the product at the time it is held in inventory. When the product is sitting in the IDC Alliance Fort Worth Distribution Center, its value is a combination of purchase price plus any transportation costs to get it from the supplier to the DC plus in-transit carrying costs. What is the *total annual inventory carrying cost* (in dollars) for the safety stock and cycle stock inventory held at the Alliance Fort Worth Distribution Center if we purchase everything from Dong Hai Supply? From CousinsAg?

9. Inventory carrying costs are based on the value of the product at the time it is held in inventory. When the product is sitting at IDC's Alliance Fort Worth Distribution Center, its value is a combination of purchase price plus any transportation costs to get it from the supplier to the DC plus in-transit carrying costs. *On a per-unit basis* (in dollars), what is the total inventory carrying cost for the safety stock and cycle stock inventory held at IDC's Alliance Fort Worth Distribution Center if we purchase everything from Dong Hai Supply? From CousinsAg?

10. Let's put it all together to determine the total cost of ownership. We have determined the unit price, the in-transit carrying cost, the transportation costs, and the IDC Alliance Fort Worth Distribution Center's inventory carrying cost. If we also consider the annual ordering cost, what is the *total cost of ownership per unit* (in dollars) if we purchase everything from Dong Hai Supply? From CousinsAg?

11. After you incorporate all the risk costs, which supplier is the *least total cost provider* of Schachtel Schmuggel Bannware?

12. There are additional risks that must be considered to better evaluate IDC's decision for the two supply chain choices—CousinsAg and Dong Hai Supply. Identify two additional risks that should be considered, and provide at least two realistic quantitative measures for each risk that you would use to evaluate that risk.

13. Recommend improvements to the supply chain process to reduce total landed cost.

Endnotes

1. Michael S. McDaniel, presenting "Modern High Seas Piracy" to the Propeller Club of the United States at Port of Chicago with November 2005 update, on November 20, 2000, www.cargolaw.com/presentations_pirates.html (accessed February 8, 2010).

2. International Chamber of Commerce International Maritime Bureau (IMB) Piracy and Armed Robbery Against Ships Annual Report January 1–December 31, 2009, p. 6.

3. Associated Press, "Pirate Attacks Drive Up the Cost of Shipping: Companies Face Higher Insurance Rates or Taking Longer, Expensive Routes," MSNBC, April 12, 2009, www.msnbc.msn.com/id/30180080/ (accessed February 8, 2010).

4. www.city-data.com/states/Nebraska-Labor.html (accessed February 8, 2010).

5. Exchange rate as of February 8, 2010. The instructor can update the rate by accessing www.xe.com/.

6. www.chinapage.com/road/shanghai-chengdu-railroad.htm (accessed February 8, 2010).

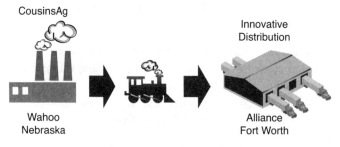

Exhibit 1 Domestic supply chain

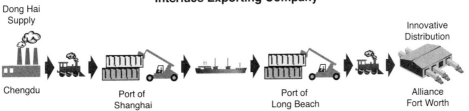

Exhibit 2　Global supply chain

10

HUMANITARIAN LOGISTICS: GETTING DONATED FOODS FROM SWITZERLAND TO ZAMBIA

David B. Vellenga, Maine Maritime and LCC International University

Case Overview/Background

The Alliance for Children Everywhere (ACE) is a faith-based organization that has provided help to children in the United States, Central America, and Africa (in Zambia). The major focus of ACE is to rescue infants and children, restore them to health with proper nutrition and care, and return these children to their immediate families or extended families, or place them with caring adoptive families. In recent years, ACE has experienced a rapid increase in the use of its home, The House of Moses (HOM), in Lusaka, Zambia, as a result of high mortality rates of parents due to HIV/AIDS, high unemployment rates (approaching 70 percent), and pervasive poverty in the country.

ACE has provided a very high level of care for infants (including premature babies), young children (up to 24 months), and also older children who have been abused or abandoned by their parents/families—through their Zambian subsidiary Christian Alliance for Children in Zambia (CACZ). CACZ enjoys an excellent reputation in Zambia and is very well known to the local population and government officials. The staff at HOM is extremely dedicated and hardworking in this 24/7 operation. Many of the staff have very long commutes to work, often one or two hours each way. Their salaries are relatively low even by Zambian standards. Because this is a private organization and receives almost no aid from the Zambian government, sources of funding are a continual issue. CACZ relies almost entirely on donations from individuals, churches, foundations, and volunteer teams.

In early 2006, a German charitable group, Giving Hands, was able to purchase a large quantity of infant and baby formula and cereals at a very low price from a Swiss manufacturer. This purchase was made possible by funds provided by Swiss and German charitable groups. The Giving Hands group made contact with CACZ at the House of Moses in Lusaka to determine its needs for infant formula and cereals. Giving Hands was cautious in preliminary negotiations due to a prior experience of misappropriation with another charitable organization. After some discussions, it seemed like a "perfect fit" between the goals and needs of Giving Hands and CACZ in Lusaka.

As a result, two full containers of baby formula and cereal were to be shipped to HOM from Basel, Switzerland. At this point in time, CACZ recognized the need for help and expertise in managing this logistics project. Few of the full-time staff had experience in logistics. This was an opportune time to seek volunteers who could help carry out this logistical challenge. Soon afterward, ACE was able to find a volunteer with logistics and SCM experience.

Routing of Shipment

The first step in the logistics process was to select the routing and modes of transportation to be used in making the shipment from Basel, Switzerland, to Lusaka, Zambia.

Several possibilities existed. For example, British Airways had direct flights from Heathrow (LHR) Airport to Lusaka, Zambia (LUN). Cargo could then be routed from Basel (or Zurich) to LHR to LUN. Other possibilities included the use of different European Airlines (e.g., KLM, Lufthansa, Swiss International) or South African Airways. These, however, required connections in Johannesburg (JNB).

Another alternative was the use of an international freight forwarder (IFF) that would make flight arrangements either on its own aircraft or on airlines. Firms such as UPS, Federal Express, DHL, Panalpina, and Schenker International were potential IFFs. Finally, there was the use of charter aircraft. One concern was that, generally, the use of airfreight is very expensive vis-à-vis ocean/surface shipments.

The other routing alternatives involved the use of water and surface transport modes. Because Basel is on the Rhine River network, the container could be shipped via barge to a major container port in the EU such as Rotterdam or Antwerp. Alternatively, they could be moved by rail or truck to Rotterdam or Antwerp. Once the container reached the export port, the ocean trade route and carrier had to be selected. There were numerous possibilities!

However, issues beyond transportation had to be considered before reaching a decision. These included the financial, political, and physical risks and the stability of transit countries. When carrying goods through multiple countries and waterways, a major concern was to avoid those regions that had high degrees of instability. Another concern was the risk of piracy in some waters near ports that could be used to route the goods to Zambia.

Potential routings included shipments from Antwerp/Rotterdam to West African ports in Angola or South Africa (Cape Town), then via truck to Lusaka. Others included routings via the Suez Canal to East-African ports such as Mombasa in Kenya, Dar Es Salaam in Tanzania, or Durban in South Africa, then via truck to Lusaka. (*Hint:* One source of routing information for ocean voyages is the Journal of Commerce. See JOCSailings.com.)

The selected route was via barge from Basel to Antwerp, then ocean vessel from Antwerp to Dar Es Salaam, Tanzania, by way of the Mediterranean Sea and the Suez Canal, utilizing the Dutch carrier West European Container (WEC) Lines. The containers would then be transported by truck from Dar Es Salaam to Lusaka. This was deemed to be the most advantageous routing because a fairly good highway exists between these cities and there are good relations between Zambia and Tanzania in terms of cross-border transportation and trade. Because Zambia is a landlocked country, it depends on ports in other countries to export and import goods (see Exhibit 1).

Shipment Details

The manufacturer placed the baby formula/cereal in two 40-foot containers in Basel. The goods were shrink-wrapped and palletized as well. Additionally, they were sorted on the pallets by product type (e.g., infant formula up to four months of age, apple or rice cereal, etc.) and by expiration date ("must use by" date). The containers were then locked and sealed prior to trucking them to the barge line for carriage to Antwerp.

Container I consisted of four types of Adapta products (infant formula) with five different expiration dates placed on 67 pallets. Total weight of the goods was 11,413 kg and consisted of 12,853 boxes of formula. Each packet of formula was packaged in aluminum foil to protect the formula from excessive temperature changes and moisture. Adapta products were packaged in cartons that consisted of three smaller boxes (each box containing two packets). Total cartons (cases) were 4,284.

Container II consisted of one Adapta product, two Galactina Humama products, and six types of cereals with eleven different expiration dates placed on 69 pallets. Total weight was 10,263 kg and consisted of 17,440 boxes of formula and cereals. Because the cereals were of lower density than the formula, container II utilized considerably more space than container I. The Humana products were packaged in a very similar fashion to the Adapta products in container I (i.e., two packets placed in a box and then three boxes placed in a carton). The cereals were packaged in aluminum foil packets of 250–350 grams and then six were placed in each carton. Total cartons (cases) were 2,994 for formula and 1,410 for cereals totaling 4,404 cartons (see Table 1).

The containers left Basel by barge at the end of March 2006 and were transported to Antwerp. Then they were loaded aboard the Dutch vessel MSC Loretta on April 7, 2006, for the voyage (# 614) to Dar Es Salaam. The estimated time of arrival (ETA) was May 7, 2006, for a voyage totaling 30 days. The containers were then processed though the port and customs, and then placed on trucks for the final stage to Lusaka. They arrived in Lusaka on May 29, 2006, and were impounded in the customs area.

As mentioned earlier, the total costs of shipping were assumed by the German charitable group, Giving Hands. In addition, they provided an allowance to pay for local charges in Lusaka—such as customs fees, local drayage, hiring of day laborers to load and unload containers and trucks, fuel costs for the House of Moses truck, and rental of a container handling forklift truck.

Because the volunteers assigned to handle the final steps of the shipment were not scheduled to arrive in Lusaka until June 20, 2006, the recipients of the donated goods were not in a hurry to have the goods released by customs.

Transportation Documentation

International shipments, especially perishable food products, require considerable documentation. In addition, additional forms are necessary in the case of donated goods in order to reduce or eliminate customs fees and minimize the possibilities for these goods to reach the underground markets for resale. The following is a brief delineation of the key paperwork/forms used in this case:

- **Shipping Bills of Lading (B/L)**—May be for the total intermodal shipment (barge, vessel, and truck) or require a separate B/L for each mode or carrier. Also needed to track and trace the shipment while en route (see Exhibit 2).

- **Shipping Advice or Notices**—Provides the shipper with details about the shipment (e.g., delivery dates, weights, packaging) (see Exhibit 3).

- **Certificate of Donation**—Proves that the goods were actually given to the recipient. Especially for relief/humanitarian goods (see Exhibit 4).

- **Packing List**—Provides detailed information on the contents of each package in the shipment. Also very useful for customs' purposes.

- **Sanitary Inspection Certificate for Food**—Is often required for both country of origin and destination.

- **Veterinarian Certificate**—Also certifies the sanitary conditions of shipments of plants, animals, and foods.
- **Use of International Commercial Terms (INCOTERMS)**—Specifies the responsibilities of the buyer and seller in an international transaction. In this case, the containers were shipped using the delivered duty unpaid (DDU) INCOTERM in Lusaka. This means that Giving Hands paid for all the shipping costs to the customs area in Lusaka. The HOM was then responsible to pay customs fees in order for the containers to be released to them.

Receipt of Goods in Lusaka

The logistics volunteers arrived in Lusaka on June 21, 2006. On June 23, HOM representatives and a logistics volunteer went to the office of a customs house broker (CHB) to make final arrangements to get the two containers released from the Lusaka customs area. HOM reps had worked with customs officials earlier about details related to the release of the goods. Initially, customs required an "expediting fee" to release the goods.

The use of the CHB resulted in a smoother process and the necessary fees were paid. Total customs and CHB charges were 4,151,000 kwacha ($1,220 USD).

The HOM wanted to purchase one of the two FEU shipping containers to use as a secure storage unit at one of its facilities in Kanyama. The cost for a used container was approximately $1,400. Therefore, upon arrival, customs container II was still securely locked and sealed. This was the container that HOM intended to purchase.

However, container I had been opened and the goods were moved to a warehouse. This was done to avoid extra charges for keeping a container for an excessive time period (May 29 to June 23). Unfortunately, this goods transfer had been done in a very haphazard manner. Cartons were tossed in piles on a dirt floor. Additionally, it was very dusty, making it difficult to breathe. To compound matters, all the pallets had been broken down and all product lines were mixed up and expiration dates were mixed as well. Control over the contents of container I was also lost. It would have been extremely difficult and time consuming to determine if all 4,284 cases were in the warehouse and if any theft or loss had taken place.

Transportation from Customs to the House of Moses

The next step in the logistics flow was to move the goods to the HOM, a journey of about 15 km. The HOM owned a Volvo truck, which was used for this movement of goods. The HOM crew consisted of a driver and three helpers. As loading began, it soon became apparent that more help was needed to transfer the cartons from the warehouse to the truck in a timely fashion. More problems surfaced as the HOM truck did not have high enough sideboards nor did they have a tarpaulin to cover the cargo in the truck bed. Cartons could, therefore, only be stacked three high, which made for very inefficient use of the truck—considering that more than 4,200 cartons had to be moved.

About four day laborers were then hired to expedite the loading and, later, the unloading process at HOM. These young men were hired for 50,000 kwacha (about $15) per day plus lunch. Most were eager to work, but didn't seem interested in trying to sort the products by type or due dates at HOM.

Finally, all the cartons from container I were delivered to the HOM. At this point, it was necessary to re-sort all the cartons by product line (four products) and five must-use dates. Volunteers at the HOM, HOM employees, and the day hires managed to accomplish this very tedious task.

It was not possible for the logistics coordinator to be at both the origin and destination for this transfer, leading to a further loss of control. Because the HOM is a secure, walled location with a gate guard, it ensured security of the goods when they were finally delivered there.

Next steps were taken to plan the move of goods in container II as well as the container itself. Attempts were made to use the lessons learned from the previous transfer of goods from container I. It should have been an easier move because the goods remained in a sealed container. Initially, it was hoped that container II could be moved directly to HOM from the customs area. However, this was made impossible by the lack of a movable crane capable of handling a fully loaded FEU container and a larger truck to transport the container. All movable cranes were hired out at the time and forklift trucks were unavailable to move the pallets from the container to the truck. Instead, the driver and crew of the HOM Volvo truck were briefed on the most efficient way to move goods from container II to HOM. Maintaining pallet integrity and not mixing product lines was stressed.

The day began well with the first delivery consisting of cartons of Adapta 2 with a single expiration date. These cartons were easily added to the Adapta 2 cartons from the prior delivery (with the must-use date of December 22, 2006). Soon thereafter, things began to unravel as products were again mixed up and little attention was given to must-use dates. Again, the driver and crew were briefed on proper procedures while delivering to the HOM. Unfortunately, the logistics advisor could not be at both the origin (customs area) and the destination (HOM) at the same time to ensure procedures were followed during pickup.

As the day progressed, it became clear that the total transfer of goods from container II could not be completed by the 5:00 p.m. deadline—the time when the customs area closed. If goods remained beyond 5:00 p.m., an additional day's storage charge would apply. Therefore, the HOM driver was authorized to "hire" another truck. This is not an uncommon practice in Zambia. One merely flags down a truck and negotiates for its hire on the spot. This was done for a total cost (including fuel) of about 500,000 kwacha (about $150). It also required the hiring of four additional day laborers at the usual 50,000 kwacha rate per day plus lunch. With these arrangements, the contents of container II were delivered to HOM by the end of the day.

However, the HOM crew members, day laborers, and the rental truck crew again produced some confusion. Boxes were "thrown" into the HOM compound being used as the staging and consolidation facility until final distribution. Some day laborers in the human chain could not keep up the pace and many cartons fell to the ground, but the foods were well packaged and were not easily damaged in the rough handling. The more than 4,000 cartons had to be re-sorted one by one—again a very time-consuming task, especially aggravated by the fact that the majority of volunteers who had helped in the previous re-sorting process were no longer there.

The next morning, as the few remaining people at the HOM were eating breakfast, a minivan arrived in the driveway. Out stepped a dozen American college students. They were attending a national conference in Lusaka. Somehow they had heard about HOM and its work with orphans and infants. They wanted to see the babies and nursery firsthand and learn more about the work being accomplished there. Next, they stated that they had no additional plans for the day and wondered if there were any projects at HOM that they could help accomplish. After visiting the nursery, they were ready to help sort out the cartons piled up in the front yard. They quickly understood the requirements and pitched in alongside the HOM employees. By the end of the day, all the cartons from container II were properly sorted and combined with container I cartons. The first phase of the project, getting the goods from the customs area to the HOM compound, was now completed.

Allocation of Foods to Other Aid Organizations

Early on in this process, it was clear that two containers of baby formula and cereal would be too much for HOM alone. It was necessary to find other aid/relief agencies that could benefit from this baby food shipment. It was also important to find organizations that had good reputations because it was not uncommon for past donated relief/humanitarian goods to turn up in the marketplace for resale.

Based on the experiences of HOM leaders, the reputations of other organizations, and prior working relationships, three organizations were selected. One was the well-known relief organization World Vision. The second was the University Teaching Hospital (UTH), a medical school affiliated with the University of Zambia, which experiences a very high demand for maternity and infant care—including premature babies. The third was the Prison Fellowship group. In Zambia, women prisoners take their younger children to jail with them. If pregnant, they give birth to their babies in prison. However, because the Zambian prison system does not allocate funding for food for these infants and children in their annual budget, mothers must somehow obtain food for their children. Some have relatives bring food to the prison, but others do not have this option.

Some small quantities of products were also given to some local groups and ministries that have an ongoing relationship with HOM: the social welfare department in Zambia, which closely works with HOM on adoption placements (100 cartons); Action International Ministry (30 cartons); Jesus Army Church and School (30 cartons); and the local leprosy home (50 cartons).

World Vision representatives came to HOM with their own truck and picked up their allocation of products in three types of formula (about 1,500 cartons) for delivery to their central store location in Lusaka. The World Vision driver fully understood the need to keep product lines together when loading the vehicle. He kept close watch on the loading process to ensure proper sorting. However, this level of micromanagement led to some conflicts with the men loading the truck, with one man walking off the job due to being "corrected." He stressed to his crew the need to keep items segregated as they were being loaded and unloaded at the central store location.

The University Teaching Hospital sent a vehicle to collect its allotment (about 2,700 cartons). The Prison Fellowship did not have a vehicle, so it hired the HOM truck to deliver its goods (about 750 cartons) to its central warehouse in Ngola, about 300 km north of Lusaka. HOM received 300,000 kwacha ($90 USD) for this transport service.

Goods Turnover Program

The HOM, Giving Hands, and other aid agencies wanted to inform the Lusaka community about this generous humanitarian aid—especially because it was aimed at helping babies and infants. The targeted demographic is severely affected by the high levels of HIV/AIDS in the country. HOM hosted a turnover program in which donors and recipients of the baby foods were present. The HOM yard was used to display all of the cartons of donated food, providing a clear picture of how much food had been donated to the Lusaka community.

The event was widely covered by Lusaka television, radio, and newspapers. Representatives from the HOM board, World Vision, the director of the University Teaching Hospital, and the head of the Prison Fellowship were there alongside the directors of the Children's Hunger Relief Fund (the major donor) and Giving Hands, and representatives from the German Embassy.

Several local pastors were present and expressed thanks for the baby foods. Women and children from the local community were also invited to a luncheon hosted by HOM. All mothers received a locally made blanket. More than 50 mothers attended the program. Overall, this was a very successful event from humanitarian and public relations viewpoints.

Repositioning of Container II

As mentioned earlier, HOM purchased container II to use as a secured storage unit at a home for older children who have been rescued from abuse or abandonment. It is about 15 km from HOM. This project required hiring a large movable crane and a large truck that could move the FEU container from customs to Kanyama. Total costs for this project were 3,791,000 kwacha ($1,115) while the cost of the used container was $1,400.

Concluding Activities of the Logistics Project

The goods retained for use by the HOM were then moved from the yard into a storage container that was already there. This involved close supervision to ensure that the product lines were properly segregated and must-use dates were visible on the front when stacked. A diagram with the precise location of each type of product and due dates was given to the HOM executive director.

Procedures were outlined and implemented to ensure security and maintain an accurate inventory record for each of the product lines. The container was kept locked to minimize the potential for theft and key custody controls were put in place.

Estimates of the cost savings to HOM for using the donated baby formula and cereals instead of those formerly purchased in the local market was $36,000 in 2006. It was hoped that these savings could allow a modest pay increase to HOM employees in the future.

There was concern about how the babies and infants would react or adjust to the new types of formula and cereals. Luckily, they quickly adapted to the new products causing no problems.

Some of the donated products had relatively short must-use dates (e.g., August 7, 2006). This was of some concern because Zambians typically discard items on the expiration date printed on the product, in contrast to many North Americans and Europeans who feel that these dates are set very conservatively—products are sometimes used for a few months beyond the due date. Again, cultural differences must be considered here.

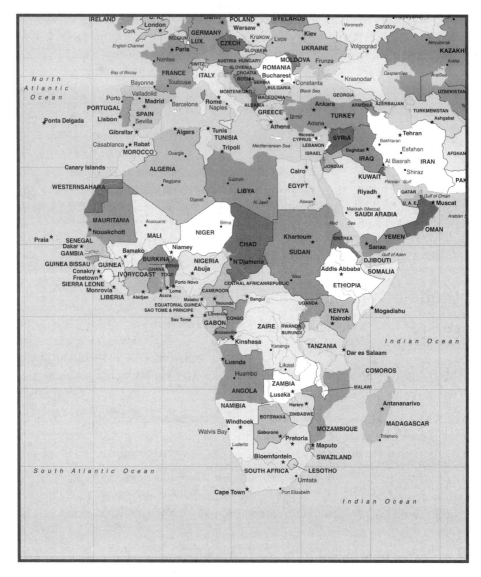

Exhibit 1   Map of Africa and the Southern EU

Table 1 Total Logistics Costs

Customs clearing/CHB	4,151,000 kwacha	$1,220
Other logistics costs [fuel, day laborers, lunches, rental truck]	1,480,000 kwacha	$435
Forklift, container move	3,391,000 kwacha	$1,115
Totals	9,422,000 kwacha	$2,770
Purchase of used container		$1,400
		$4,170

Exchange rate used is 3,400 kwacha/USD.
Note: Exchange rates have fluctuated in the last 18 months from 3,100 to 5,600 kwacha/USD. This makes budgeting and cost control efforts very difficult.

Bill of lading for combined transport or port to port shipment

Shipper: HERO, NIEDERLENZER KIRCHWEG 6

Consignee: CHRISTIAN ALLIANCE FOR CHILDREN IN ZAMBIA

Notify Party: *** CTND CNEE:

B/L No. WECC0601LUS0306
Key 58502
Reference 3659/6467

Original 2

W.E.C. LINES — WEST EUROPEAN CONTAINER LINES

Combined transport (subject clauses 6 and 7)

Pre-Carriage by	Place of Receipt
Ocean Vessel	Port of Loading: ANTWERPEN
Port of Discharge: DAR ES SALAAM /TANZANIA	Place of Delivery: LUSAKA

Freight Payable at: ROTTERDAM
Number of original Bs/L: 3

Marks and Numbers / Containers nos. & seals nos.	Number and kind of packages	Description of goods	Gross weight in kilos	Measurement in cbm
//// 408124 1 SEAL 1008530	= 2 X 40 BOX 12953	SAID TO CONTAIN PKGS. RELIEFGOODS/HUMANIT. GOODS ACC. LOADING SPECIFICATION	11.417,00	
CLHU 417154 9 SEAL 1008528	17440	PKGS. RELIEFGOODS/HUMANIT. GOODS ACC. LOADING SPECIFICATION	10.259,00	

RELIEFGOODS /HUMANITARIANGOODS
DELIVERY DUTY UNPAID LUSAKA
NO COMMERCIAL VALUE
DONATED RELIEFGOODS ACC. ENCL. CERTIFICATE

PARTICULARS AS DECLARED BY SHIPPER

FREIGHT PREPAID
Shippers load stowage and count

FREIGHT AND CHARGES	Mode of transport: FCL. / FCL.

Received by the Carrier the Goods as specified above in apparent good order and condition unless otherwise stated, to be transported to such place as agreed, authorized or permitted herein and subject to all the terms and conditions appearing on the front and reverse of this Bill of Lading to which the Merchant agrees by accepting this Bill of Lading any local privileges and customs notwithstanding.
The particulars given above as stated by the shipper and the weight, measure, quantity, condition, contents and value of the Goods are unknown to the Carrier.
In witness whereof one (1) Bill of Lading has been signed if not otherwise stated above, the same being accomplished the other(s), if any, to be void. If required by the Carrier one (1) original Bill of Lading must be surrendered duly endorsed.

SHIPPED ON BOARD
Place and date of issue: ROTTERDAM 08/04/2006
Signed on behalf of the Carrier: West European Container Lines

Total no. of Packages	DEMURRAGE

Appendix A

Exhibit 2 Bill of lading

CHRISTIAN ALLIANCE FOR CHILDREN IN ZAMBIA (CACZ)

12 APR 2006

SHIPPING ADVICE

ON BE HALF HERO
    NIEDERLENZER KIRCHWEG 6

WE HAVE SHIPPED THE FOLLOWING SHIPMENT TO THE PORT OF DAR ES SALAM/TANZANIA

SHIPMENT   -   XXXX 408124-1 1 X 40 FT SHIPPERSOWNED
CONTAINER S.T.C. TOTALLY 12853 PKGS.
RELIEF/HUMANITARIANGOODS 11417
KOS ACC. ENCL. CERTIFICATE OF DONATION AND
PACKINGLIST.
CLHU-417154-9 1 X 40 FT LINEROWNED
CONTAINER S.T.C. TOTALLY 17440 PKGS.
RELIEF/HUMANITARIANGOODS 10259
KOS ACC. ENCL. CERTIFICATE OF DONATION AND
PACKINGLIST.

TERMS OF DELIVERY   -   DELIVERY DUTY UNPAID LUSAKA

VESSEL   -   MV MSC

E.T.S. ANTWERPEN   -   7 APR 2006
E.T.A. DAR ES SALAM   -   7 MAY 2006
E.T.A. LUSAKA   -   22 MAY 2006

ENCLOSED DOCUMENTS   -   CERTIFICATE OF DONATION
PROFORMA INVOICE NR. 3659
PACKINGLIST
SANITARY INSPECTION CERTIFICATE FOOD
2/3 ORG BL AND 2 NN COPIES
2 X ORIGINAL VETERINAIRCERTIFICATES
2 SETS OF CERTIFICATE OF ANALISE
AMENDMENT OF HERO/VETERINAIR

PLEASE CONTACT AGENT AS SOON AS POSSIBLE:
Name Tanzania Shipping Agency Ltd.
Address Off Nelson Mandela Road
  Kurasini Road

PLEASE LET US KNOW SOONEST IF SOMETHING IS NOT TOTALLY
CLEAR.PLEASE ARRANGE IMPORTLICENSE ASAP.

THANK YOU FOR ORDER

WITH WARMEST REGARDS,

JAN DEURLOO
MISSION & RELIEF TRANSPORT
ZEVENHUIZEN

Appendix B

Exhibit 3   Shipping notice

STOP HUNGER NOW

www.StopHungerNow.org

Letter of Donation

Date: 4/17/09

To: Customs Officials of ZAMBIA and
    Whomever Else it May Concern

This letter is to certify that this shipment of DONATED RELIEF GOODS: dehydrated rice casseroles, donated shoes, and donated medical supplies is being sent through STOP HUNGER NOW as a free gift to the people of ZAMBIA. The consignee, who is responsible for handling the shipment, is:

Christian Alliance for Children in Zambia (CACZ)
c/o House of Moses

Lusaka, Zambia
Contact Alice

ORIGINAL

This shipment is to be administered by CHRISTIAN ALLIANCE FOR CHILDREN IN ZAMBIA and the relief agencies and mission groups working with the consignee. The contents of the shipment are to be used for HUMANITARIAN PURPOSES ONLY. The contents may be either distributed directly to the needy or used to manage and set up the logistical needs of its aid programs. The shipment is not to be sold, resold, or exchanged for profit or gain. Therefore, there is no commercial value to this shipment. The declaration of value, US $10,000, is for customs purposes only, and does not involve any currency of ZAMBIA. CHRISTIAN ALLIANCE FOR CHILDREN IN ZAMBIA is hereby given permission to administer this shipment in the manner that it finds to be the most beneficial to the poor an needy peoples served by its mission and aid programs. This includes the sharing and re-donation of the contents of the shipment to other relief and development agencies.

Accordingly, it is requested that those parties handling the receipt, clearance, and onward forwarding of this shipment process it expeditiously and in good faith, so that the relief and charity efforts in ZAMBIA can begin as soon as possible. Any changes to this statement needed to comply with the local rules and regulations may be made if in agreement with and attested by the signature of the consignee.

Sincerely,

Melissa Holmes
For STOP HUNGER NOW

**DONATION DONACION
Not For Sale No Para Venda**

Appendix C Illustration only: This document is not from the case shipment

Exhibit 4   Donation letter

Part 4
THE FUNCTIONS

This section contains three cases: (1) "Breaking Ground in Services Purchasing," (2) "Lean at Kramer Sports," and (3) "UPS Logistics and to Move Toward 4PL—or Not?" Although the focus of the three cases is best described as addressing several of the basic, functional tasks within the supply chain, each case examines a particular functional task and its attendant challenges for managers.

The "Breaking Ground in Services Purchasing" case focuses on an area of organizational spending (i.e., purchasing) and the improvement of efficiency and effectiveness in services procurement. The case illustrates the utilization of category management and purchasing segmentation to the purchase of services. It demonstrates the subtle and not-so-subtle differences in the purchasing of goods/materials versus nontraditional services. The case also provides for consideration of supplier performance evaluation, and it suggests that a component of such evaluation will likely lead to consideration of supplier relationship management.

The "Lean at Kramer Sports" case focuses on the well-known concept of lean but with a very important and interesting "twist." The assumption that you can only learn from success is one-dimensional thinking and this case offers an opportunity to see "the other side of the coin:" a company embracing mass production principles while failing to benefit from the "magic wand" that lean often promises. The case demonstrates a company that attempted and failed to convert to a lean organization and ultimately lost in bigger ways. The case also illustrates that myopic thinking prevents change, managers/strong leadership/adaptive culture are a key element to change and success, and that sometimes a unique organizational structure does a company no favors in converting to lean.

The "UPS Logistics and to Move Toward 4PL—or Not?" case focuses on the concept of fourth-party logistics (4PL) and related strategy formulation, particularly with regard to full supply chain integration. The case demonstrates the time and effort required to achieve a successful 4PL client relationship, and then what challenges likely will lie ahead to expand that newly created business model. In particular, the case introduces third-party logistics (3PL) business management considerations with regard to outsourcing trends, industry challenges, and manufacturer (i.e., customer/client) opportunities and perspectives. It also points out a 4PL business model will result in the company competing with different companies than it is used to.

11

BREAKING GROUND IN SERVICES PURCHASING

Lisa M. Ellram, Miami University

This case was prepared for classroom discussion purposes only. It is based on a real scenario. Some of the data and verbiage from conversations have been modified to enrich classroom discussion.

Background

Gensin is a global, *FORTUNE* 100 company in the pharmaceutical industry. It experienced considerable growth due to both acquisitions and successful new product introductions. While its decentralized model had been successful in many regards, it was also very expensive to operate. In addition, Gensin had a desire to have a stronger presence with its customers. A decentralized model did not easily allow Gensin to put "one face forward" to the customer. Thus, Gensin was going through some significant reorganization that would support both improved cost efficiency and more effective customer service.

Alice May had just taken over the newly created worldwide purchasing organization at Gensin. Although May had nearly 20 years of experience leading purchasing organizations—including nearly a decade at Gensin—a centralized, strategic purchasing organization was a new concept at Gensin. Prior to this reorganization, the purchasing organization had been decentralized, organized around business units, and loosely coordinated through a global purchasing council, as shown in Exhibit 1.

Rather than organizing around specific lines of business, Gensin was changing its structure to organize around the customer, as illustrated in Exhibit 2, on the new customer organization chart. The issues guiding the new organizational design included a number of factors. First, there was a desire to better serve emerging markets, which were growing at a much faster rate than the more established markets in developed countries. The emerging markets had much lower price points than established markets. The needs of these markets could be best met by production from low-cost, third-party manufacturers, which were often located closer to each particular market. The production of these third parties would cut across businesses and could be best coordinated centrally.

The combination of a globally expanding customer base and a global, third-party manufacturing base created an increasingly complex global supply chain. To improve supply chain responsiveness and information transparency, central coordination was more efficient. The focused business units shown across the top of Exhibit 2 would be able to stay abreast of dynamic changes in product and customer needs, and work with the centralized, customer-focused support organizations in a more efficient manner.

Purchasing Reorganization

Thus, the purchasing function was reorganized in order to provide more coordination across businesses and with other functions and to provide a greater focus on the end customer, in line with the reorganization of the businesses. The new organization for the purchasing area is shown in Exhibit 3.

The principles guiding the reorganization of purchasing included the following:

- Category management that continues to focus on spend and supplier synergies. This approach leverages Gensin's global scale operational support, aligned with the customer, developing a flexible/practical model—not one that is "one size fits all."
- Future governance that focuses on connecting with the internal customer and less on internal purchasing consensus-building of decentralized purchasing personnel.
- Effective distribution of resources, including more category management resources outside the United States.
- Reduced operating costs of the new model.

As part of this reorganization, a new role of business relationship manager was created. The role is to do the following:

- Be fully aware of customer strategic business needs and identify supporting opportunities for worldwide purchasing.
- Serve as a senior, single point of contact for internal customer organizations.
- Coordinate with other units of worldwide purchasing, as necessary, in support of the customer.

Category Management

In addition, given the new organization and greater visibility within purchasing, May and her team began to methodically implement category management. The old purchasing organization used category management; however, it was done informally and relied on purchasing councils of people with purchasing responsibilities. These people were decentralized (spread among the business units), had varying responsibilities, and reported to various levels and even different functional areas. The new centralized structure created a great springboard for improvement in category management.

Just as marketing groups similar products and services that it sells into categories for marketing, management, and positioning to customers, purchasing category management represents a way to group like purchases of goods or services into categories. The purpose of doing this in purchasing is to analyze the spend patterns; to identify methodical ways to leverage the purchases of like items; and to get better quality, service, and pricing. Just as marketing needs to understand its product offerings, avoid duplication, and understand why each item it sells is unique, purchasing must examine each item it buys for possible synergy, duplication, and best value. Without coordinating similar purchases of like items, Gensin was spending too much time and effort managing too many suppliers, duplicating efforts and processes. In addition, Gensin was paying too much because it did not leverage its volume and, consequently, did not appear to be as important a customer to key suppliers.

Exhibit 4 shows Gensin's "as is" and "desired" states regarding purchasing. Gensin wanted to put "one face forward" to suppliers, reduce waste, and work with best-in-class suppliers in every category.

Gensin developed a set of sourcing guidelines to guide it every step of the way. These were developed by benchmarking best practices of other companies, leveraging relationships in supply management professional organizations (such as CAPS Research and the Institute for Supply Management), and working with top consultants in the procurement profession. These final sourcing guidelines included the following approaches:

1. Rationalize the supply base. In most cases, it meant reducing the number of suppliers, while in others it meant adding new and different suppliers who could better meet Gensin's needs.

2. Leverage Gensin's spend with suppliers. This would help make Gensin more visible as an important customer, wherever possible.

3. Evaluate the total cost of ownership. This was in respect to each purchase and involved working with a supplier or category, rather than focusing just on low price.

4. Employ a global supply base. This helped Gensin utilize the best suppliers throughout the world.

5. Include all internal customers. This included involving those individuals who do any buying with purchasing as early in the buying process as possible, so that purchasing can add value instead of just performing a signatory function.

6. Use preferred purchasing methods appropriate to the category. This included e-auctions, contractual relationships, procurement cards, and other up-to-date methods.

7. Mandate Gensin's use of purchasing's preferred supplier agreements. This is the case when such agreements are in place.

One key issue to be kept in mind is that the overarching goal of category management is to achieve and support the business objectives, not just to save money. With these guidelines in mind, the steps that Gensin followed in implementing category management are detailed in the graph shown in Exhibit 5.

These steps were verified by benchmarking best practices, reading the latest research, and conferring with consultants. Category management implementation began in traditional categories, such as packaging and materials, or ingredients—areas where purchasing already had a very high level of involvement—where its participation was expected and its value-add was clear. While there was a great deal of savings to be gleaned from such categories, purchasing was also very interested in getting involved in nontraditional categories where it often had limited, if any, input. These categories, such as advertising, clinical research engagement, and professional services, were often handled almost exclusively within the functional area, with purchasing only stepping in to review contracts and sign off. The value-add proposition by the purchasing area in these categories had been minimal to date. Because these categories represented a huge amount of the budgetary spend of a particular area, and because the budget holder in the area generally viewed itself as an expert, these areas were somewhat politically sensitive. Purchasing would have to "sell" its value to convince the marketing group, for example, to let it perform a category analysis on advertising spend. Still, Gensin was going through a significant review of all its spending and budget reductions in many areas, so there was a greater opportunity for purchasing involvement in nontraditional areas.

One key strategy was to look at the value each category represented to the organization and compare that with the risk or complexity of the supply market for that particular category, as represented in Exhibit 6. Using the category management technique allowed the purchasing organization to focus its energy on effort in areas where it could truly add value while keeping in mind the key sourcing guidelines.[1]

Legal Spending

May was somewhat surprised, but also elated, when she got a call from Conrad Cole, the new head of the Gensin legal group. He said, "Alice, I have reviewed the way that we are buying our external legal services here, and I think that there is a big opportunity for improvement. Right now, we are working with hundreds of law firms in the litigation area. This is creating a high potential for duplication and inconsistent resolution of issues across regions. There is limited, if any, coordination between these firms, which means we may be missing opportunities to achieve the best outcomes for Gensin, and may have inconsistent outcomes across different regions. In addition, there's a lack of consistency and transparency in the way law firms determine and apply their rates. I am going to call a meeting of our top law firms and tell them that Gensin wants to change the way that we do business with them to address these and other issues. As the head of our purchasing organization, you can help us restructure these relationships. I want you to attend this initial meeting and be part of this effort."

May was thrilled by this opportunity. Purchasing had little meaningful involvement with the Gensin legal group up to this point. The nature of purchasing's involvement was primarily clerical, approving agreements that legal had already made and, in some cases, answering simple procurement-oriented questions. With more than $300 million[2] in litigation spend on the line, this represented a tremendous potential area of involvement for purchasing, where the benefits to Gensin could be very significant and very visible. This also represented a potential opportunity for purchasing to break into meaningful involvement in other areas of legal spending and other areas of the company. May knew that this opportunity had to be embraced, but she also knew that it had to be approached in a professional and skilled way that would honor the knowledge and expertise of the Gensin legal group, while still allowing purchasing to be engaged and make a contribution. As she saw it, the major issues driving this opportunity for purchasing involvement in legal spend were as follows:

- Gensin Litigation Group was working with hundreds of law firms
- High potential for duplication of work across regions
- Limited coordination between firms
- Law firms charging inconsistent hourly rates
- Lack of transparency regarding how hourly rates are established by law firms
- Desire to reduce spending at Gensin in all areas

To prepare for her meeting with Conrad Cole, May decided to look at the various spend categories within Gensin Legal to get an understanding of where they could have the biggest impact. Table 1 shows the major categories and the amount of spend for Gensin's Legal Department.

Meeting with Legal Department

With a solid overview of each of the legal spend categories, May felt as prepared as she could be for her meeting with the new head of the Legal Department. She felt confident that she and her staff would be able to help them analyze spend, but felt less confident about their actual expertise in the subtleties of legal spending, the goals of the department, and other very specific issues. She knew that if the Legal Department wanted purchasing's support, her group would have to form a very close working relationship and rely heavily on legal staff to guide them through the legal processes, characteristics, relationships, and other intricacies. She was prepared to present the case to Conrad Cole.

When she met with Conrad, she realized he already had his own case prepared. "As you know, we really need your help. Although we have many excellent outside law firms that we engage, things have become unmanageable. We are just using too many players, and they are not coordinating their efforts. I'm glad you're on board, Alice. I believe we have a tremendous opportunity here to simultaneously improve performance and save Gensin a great deal of money. Here is what I see. We have far too many firms charging rates that are all over the place. There are so many of them and the fee schedules so complex that my inside counsel ends up spending too much time trying to make sense out of billing and supporting the accounts payable staff when they need to be working on strategy. In addition, I am not sure if we really know which of these firms is best in each of the areas because we don't have a formal evaluation mechanism. Although some outside firms may lose big cases at times, it is hard to assess whether they still arranged a good settlement under the circumstances. Think about this for a while, and we can get back together before I meet with a group of the leading firms to tell them about our plans for consolidating and changing the management of the supply base. I am arranging the meeting about a month from now. I want you to be there, and we will present our case together."

Discussion Questions

1. Apply the seven steps that Gensin follows in implementing category management mentioned in the case to the area of litigation, within the category of legal spend.

    Explicitly address the following issues:

    a. Who should be on the cross-functional team, and why?

    b. What sort of data do you need to collect both internally and externally?

    c. How can the data be obtained?

    d. What should some of the key considerations in supplier selection be? Which criteria are the most important?

    e. What should be considered in ongoing monitoring and measurement, and who should do this?

2. Apply Kraljic's Matrix (preferred purchasing matrix) for classifying commodities and classify the areas of legal spend into the quadrants. Based on this classification, which should be the next several commodities for purchasing involvement, and why?

3. Develop a plan and prepare an outline to present at the follow-up meeting.

4. Discuss the benefits of purchasing's involvement in legal spend at Gensin.

5. Discuss some facilitators and barriers of purchasing's involvement in legal spend.

6. Consider some other (nonlegal) areas of spend where purchasing could play a role.

Endnotes

1. Developed based on Peter Kraljic's article "Purchasing Must Become Supply Management," *Harvard Business Review* (Sept.-Oct. 1983): 109–117.

2. This value is fictitious and meant only to represent a level of magnitude. This is not based on the actual amount of money spent on litigation for the case study firm.

3. These numbers are fictional; they do not represent the spend per category of the company on which this Gensin case study is based. But the relative magnitude and ranking of the numbers is still meaningful.

4. There is an additional, approximate $150 million spend on litigation-related expenses not managed under this umbrella, due to a highly specialized nature.

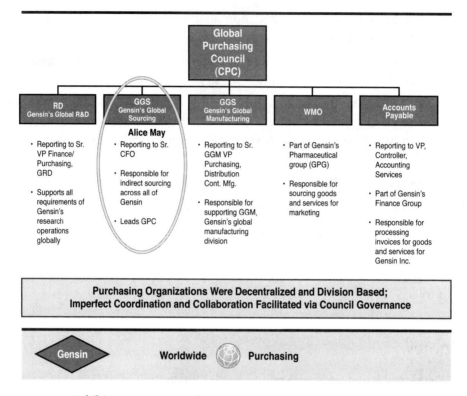

Exhibit 1 Gensin Purchasing: current organization structure

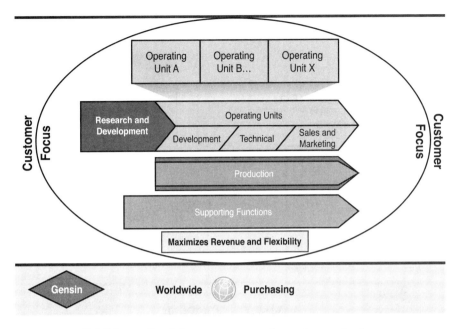

Exhibit 2 Gensin's new customer-focused organization

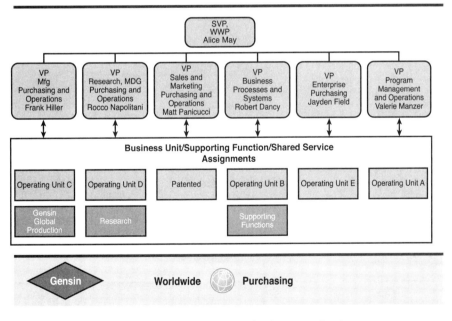

Exhibit 3 Gensin: new purchasing organization

Case 11 Breaking Ground in Services Purchasing

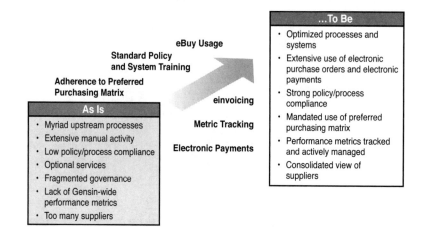

Exhibit 4 Gensin's purchasing vision

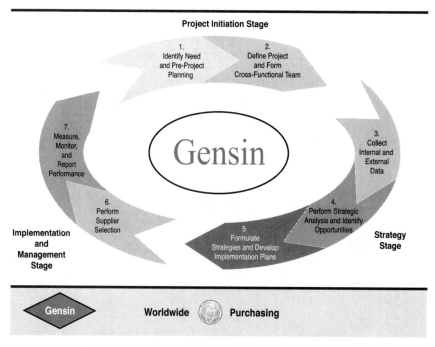

Exhibit 5 Steps in implementing category management

158 THE DEFINITIVE GUIDE TO SUPPLY CHAIN BEST PRACTICES

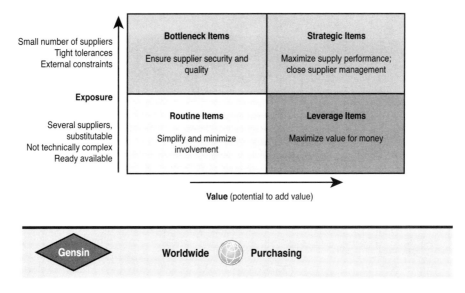

Adapted from Kraljic "Purchasing Must Become Supply Management," *Harvard Business Review* (Sept-Oct 1983), 109–117.

Exhibit 6 Preferred purchasing matrix

Table 1 Spending Summary

Legal Category	Annual Spend Millions[3]	Description	Market for Category	Exposure, Risk, Complexity
Intellectual Property	(IP)$40	This area of law protects and enforces copyrights, patents, and trademarks.	Intellectual property law provides protection and enforcement of patents, trademarks, and copyrights. Protection of intellectual property is central to any creative endeavor (such as drug discovery) that requires the commitment of substantial amounts of money. Pharmaceuticals have one of the lowest, effective patent life cycles of any product (approximately 11–12 years). Prosecution and filing of patents occurs globally.	The risk of intellectual property infringement is increasing because of the Internet and globalization, with differential legal enforcement. Infringements on branded intellectual property, trademarks, and copyrights are not easy to detect and IP protection is more complicated. Patent law is a complicated area of law governed by a confusing set of statutes and regulations. IP protection is not consistently enforced in international markets. There is inconsistency in rulings over time. High investment is required to develop intellectual property. The time-consuming research process only allows for a short period of time during which Gensin can sell products without competitors producing an exact copy.

Legal Category	Annual Spend Millions[3]	Description	Market for Category	Exposure, Risk, Complexity
Litigation	$300[4]	Litigation resolves disputes using the due process of the law. It includes trying the cases in a court of law in front of a judge and jury.	Gensin is party to a number of lawsuits stemming from companies with which it has merged or has acquired. All pharmaceutical companies are subject to tort litigation. Potential liability from drug trials and product use exists.	Spend in this area is expected to increase because of economic, technological, and environmental change—health care is expected to see significant increases in filings. Legal litigation is highly complex with each case requiring different levels of preparation and highly skilled orators. Select a firm for its attorneys (not the firm per se), their trial with jury expertise, and their track record of wins with key cases and clients.
Compliance	$25	This area of law ensures that an organization follows relevant laws, regulations, and rules aimed at business, including adherence to ethical codes.	The corporate compliance program represents the highest levels of management to the most junior employees. The company compliance program follows guidelines for the Compliance Program for Pharmaceutical Manufacturers. Compliance impacts sales and marketing, environmental health and safety, and other areas of the business.	There are written policies, procedures, and guidelines. Training and education programs are in place. Internal monitoring and auditing must occur. Standards are enforced through well-publicized guidelines. There is a risk assessment and mitigation plan in place specifically as it relates to product risk. Global rules, policies, and regulations are in place. The firm must ensure ethical behavior worldwide.

Table 1 Continued

Legal Category	Annual Spend Millions[3]	Description	Market for Category	Exposure, Risk, Complexity
Insurance Law	$6	This area of law covers insurance policies and claims, life and health insurance, reinsurance, claims, and disputes.	Legal counsel is retained to represent employees and the company in matters of insurance.	Counsel provides for and manages any and all matters related to insurance claims, including life and health insurance, and insurance disputes.
Discovery	$15	Discovery locates, retrieves, and produces data to discover pertinent facts.	Partner with in-house and outside counsel to manage case-specific identification, preservation, collection, analysis, processing, review, and production efforts. The partner provides project management, budget management, and process support. Manage legal discovery projects, outside service providers, and add value to decision making across a matrix organization. Process, host, review, and produce required data.	Scope of discovery includes legal matters related to litigation, compliance, government investigations, employment law, IP enforcement, antitrust, and business transactions. Risk is associated with inability to discover appropriate and necessary documentation. Discovery documents must be presented in a timely fashion to avoid fines.

Legal Category	Annual Spend Millions[3]	Description	Market for Category	Exposure, Risk, Complexity
Other Legal Spend	$35	Other areas of spend related to the Legal Department are common across major legal categories (i.e., consulting, court reporters, trial graphics, managing documents, and data).	Legal related work is performed by outside service organizations.	Many areas of spend are related to the practice of law at Gensin. Much of this spend was managed by outside counsel and billed back to Gensin.

12

LEAN AT KRAMER SPORTS

Steve Medland, Colorado State University
Susan L. Golicic, Ph.D., Colorado State University

Kramer Sports CEO Tim Wilcox slumped over his desk looking at a simple, handwritten list he made one year ago, comparing its hopeful words with a disturbing reality. Wilcox made the list when Kramer Sports first made the decision to alter its production environment to adopt the principles of lean production. Based on the highly successful Toyota Production System and being used effectively by organizations throughout the world, lean had helped others reduce inventory by 50 percent or more; cut production costs; and improve quality, customer satisfaction, and company morale. At the time, Kramer needed all of these benefits, as it faced a difficult business environment. As Wilcox reviewed his list, he recalled how Kramer had worked with a consultant for six months and for six additional months on its own. Wilcox wondered why Kramer's reality differed so greatly from his vision of one year ago.

Kramer Sports

Kramer Sports is a worker-owned cooperative based in the United States that manufactures and sells both high-quality bikes and bike trailers throughout both the United States and Europe. Founded in 1978, the company started with trailers before expanding into specialty racing, tandem, and recumbent bikes—later adding more trailer products, selling all of its products only through bicycle shops. Bike-avid employees were drawn to Kramer's laid-back, employee owned and managed operation. The company grew in success in revenues, reputation, and profit through the 1980s and 1990s. By the beginning of 2000, Kramer was selling about 30,000 of its three trailer models per year, and just under 2,000 of its high-quality specialized bikes in 14 frame shapes (6 touring frames, 4 recumbent sizes, and 4 tandem frames) all available in 10 colors.

The Employee-Owned Cooperative

As an employee-owned cooperative, each owner/employee was entitled to a share of Kramer's profits at the end of each year, paid in the form of a bonus. Some profit was retained each year for operating and cash-building purposes, but the bulk of the profits were paid out to owners. With the exception of owners in their first year of employment, every owner received the same bonus. Also, aside from a few managers or employees with advanced or professional training (CEO, CFO, HR, design engineers, and accountants), all employees received the same hourly wage (about $6 above minimum wage) regardless of time with the company or responsibilities. This fairly low, across-the-board wage meant that an owner's bonus could represent a significant portion of his or her annual income. It also meant that area supervisors and the people reporting to them made the same amount of money, giving them all equal incentive to make their areas productive. The two production department managers were on a slightly higher salary than their reports.

Decision making in the co-op was also different than most companies. Although each manager and supervisor had responsibility over a few workers in their area, most major decisions that affected policy, product changes, employment (hiring or firing), or benefits changes required a vote by either the full team of owner/employees or the board of directors. The board was made up of nine employee/owners representing various departments and elected by the owner/employees in their departments. Even the CEO had limited power when it came to policy changes and large expenditures as he or she needed to get board approval for many decisions. In reality, the CEO reported to the board, which was made up of the employees that reported to the CEO (see Exhibit 1).

Once an employee became a full owner (after one year), termination of employment could only be completed with just cause, and through a majority vote of current owners, 75 percent of which must participate in the vote. Even the person who was up for termination got a vote! This system prevented power-hungry managers from firing someone against whom they held a grudge, without agreement from a large number of other owners.

Factory Operations at Kramer

Kramer's production facility was laid out in two departments: trailers and bikes. Within each, processes were laid out functionally with multiple machines of the same type closely arranged in one area of the building, sometimes based on special electrical or water needs. Although most employees knew how to perform multiple jobs in the process, they generally became "experts" in one area that they enjoyed the most. Because each worker generally had responsibility for one part of the process, they would process a batch and then pass that batch on to the next process in line. Due to the varied nature of the work and cycle times, batch sizes were different for each process and sometimes for each worker. Batches moved in various ways: Sometimes the worker who completed his or her work would move one day's worth at the end of a shift, or maybe half a day's worth twice per day. At other times, a worker from the next process would come to the area and pick up what he or she needed if he or she were out of components needed for the station. Sometimes this was necessary as the nearly 100 owners at Kramer could set their own hours. Most worked 35 hours each week between the hours of 6:00 a.m. and 8:00 p.m. from Monday to Friday. This flexibility meant that parts didn't always get delivered when needed by the next station.

One advantage of having so many highly skilled factory "experts" in the building was that Kramer was capable of making many of its own tools. For example, Kramer made various sizes and shapes of carts for moving fabrics, wheel assemblies, and half-finished bikes and trailers for use all over the factory floor. The homemade racks used to paint and dry the bikes were capable of holding up to 20 bikes for powder coating (all the same color to prevent getting more than one color of powder on a frame due to overspray) and then could be rolled directly into the large "oven" for drying the coat into a hard, shiny finish. This oven was the last stop before the two final assembly stations.

To keep up with demand for the popular trailers, Kramer produced as many trailers as it could on a daily basis—yet still couldn't seem to keep up. Kramer made its most popular trailer model, the "Jimmy," for two full days each week and then switched to the less-popular "Reno" for two days, then finished off the week with the most complicated assembly, the "Boulder," for the full Friday shift. The finished Boulders would typically stay in Kramer's warehouse for a few weeks, while the Reno models would stay for only a week. Finished Jimmy trailers hardly hit the floor as many were sold before they even got produced. There was talk of producing the Jimmy for three days each week but the fear was that it would put all three models into "crisis mode" like the Jimmy. Besides, the current schedule allowed just one or zero model changeovers each day, a process which took almost an hour at each station. A midday changeover on Thursday or Friday would really drop productivity of the whole assembly line.

Tough Times

After years of growing demand and profits, lower-cost foreign imports appeared in large general merchandisers and sporting goods chains in the early 2000s. As the management team at Kramer struggled with this issue, profits and year-end bonuses dwindled. In fact, the 2002 bonus that had always been in the thousands came to a total of $870 and after fiscal year 2003, the board approved a dividend of $200 per owner—even though the company lost money and the bonus was paid out of accrued savings. The future looked no more promising. Managers even had discussions about downsizing in order to stay afloat. As a result of the cash shortfall, Kramer found some of its most experienced and hardest-working owners leaving for jobs with higher or more predictable pay.

Lean Implementation at Kramer

John Vaughn, the consultant hired by Kramer to help implement lean processes, started the project with training sessions that included the principles of lean production, value stream mapping, 5S, set-up time reduction, kanban and "pull" systems, and one-piece flow. Vaughn gave employees an opportunity to experience the tools through a hands-on lean simulation.

After the training sessions, John turned his attention to the factory floor where he spent time in both the bicycle and trailer sides of the business. In trailers, John focused on finding similarities and differences in the three trailer models in order to create a *takt* time for each and create a single process flow that would work for all three with as little modification as possible. He helped the trailer team create value stream maps for the individual processes in order to determine current batch sizes and processes being used. Vaughn pointed out the differences in batch sizes to the team members, and owners started to see how the benefits of large batches or optimization at one station didn't always benefit the system as a whole. The team made strong progress over the six months of John's weekly visits, reducing batch sizes in most areas and creating flow in the operation. The final assembly operation was split into two separate assembly lines, each able to change from one model to another in about 8 minutes. The supervisor and team enjoyed the challenge of improving set-up times and producing more trailers with less inventory and confusion. They even agreed on a set starting time so that assembly could be started in unison and produce consistently throughout the day.

However, the final assembly steps were preceded by large batch operations in two areas: frames, where aluminum bars were cut, bent, and punched to create the structural frame, and sewing, which made the various pieces of sturdy cloth that formed the bottom, top, and sides of the trailer. The trailer manager didn't seem to buy fully into lean. "Lean is great for some areas of the building," he believed, "but others are better off using a more time-tested approach to keep station and individual efficiency high."

The bicycle area had trouble adopting lean as the process was more elaborate and products more varied with the many frame/size/color combinations. For example, the two final assembly stations were each capable of building one bike at a time, and the preceding steps worked best using batch processes. As with trailers, the cutting of just one piece of a bike at a time would require some tool changes so a few were done at a time to avoid the wasted set-up time. Welding was able to do batches of four using a four-sided stand where the welder placed frame parts into the stand and then welded the points where the pieces met. The stand spun so that the welder remained stationary while the bicycles spun to him. There were five such stands for various sizes and types of bikes, each holding four bikes.

Another issue was powder coating. No one could find a way to coat more than one color on the rack without getting overspray on the other frames. It was possible to coat one bike and leave the rest of the rack empty and just bake one at a time. But why bake just one when the rack and oven were so big? This was certainly not efficient and would likely fall into the category of waste in the lean world. John Vaughn spent as much time in the bike area as he did in trailers, and found the owners receptive to his ideas. However, he was only there one day per week. During his weekly meetings with the group, the manager assured him that he would keep things moving forward and that breakthroughs were coming soon. Still, every week that John came by, he saw little change and in fact often had trouble talking to the same people week after week due to the varied nature of the work and schedules.

For his part, Wilcox was often on the road visiting large customers to make critical sales. When in town, Wilcox met with the two production managers to get updates and each told him lean success stories, which he shared with employee/owners and the board. The board seemed happy with the progress but Kramer was still not seeing any change in its financial state. The board didn't know how many more months they could lose money and needed the lean improvements to turn into financial improvements in weeks, not years, in order to be considered truly successful.

One-Year Update

To assess lean's progress, Wilcox asked John Vaughn to perform a one-year review, which included a tour of the factory, interviews with random floor employees and managers, and a summary report. After the tour and interviews, Vaughn spent some time compiling some quotes to show Wilcox during their meeting, and shared the following with him:

"Our department manager completely dropped the ball in that area. He failed as facilitator of the lean team. He failed to keep us moving forward. We have a manager that isn't very functional and the CEO can't replace him."

"There is a faction within the company that is resisting all of the attempts being made to trim inventory, to increase productivity, all of those things. They liked it the way it was so they were able to successfully reverse the changes."

"It's really difficult to make change happen here because the workforce enjoys such great power relative to other companies. People are pretty secure in their jobs. I think you get individual managers who are afraid of making change happen."

"If Tim Wilcox left, I would assume the project would stop tomorrow or soon thereafter unless whoever replaced Tim came in with an equal level of commitment, but I'd characterize that as unlikely."

Immediately following his meeting with Vaughn, one year after approaching the board with the lean proposal, Wilcox opened his file cabinet, and reached for a folder simply titled "lean" to find his list in the very back.

Wilcox's List

Why lean will succeed at Kramer:

- We have the perfect owner/employee structure for adopting lean. Each owner will benefit from lean through lower costs, higher quality, and higher profits and bonuses. Everyone will be committed to lean.
- Lean is a powerful set of principles and practices, yet easy to understand. It is easy to see the benefits of lower inventory.
- Like Toyota, we have always promoted from within and our managers have the respect of their departments due to their experience on the floor.
- As a small company, we can easily change our processes due to our current lack of strict process rules and flexible workforce.

Wilcox looked at his list again and verified in his mind that all were true. With all of these factors in Kramer's favor, why was lean failing to take hold, and what could be done to get it on track?

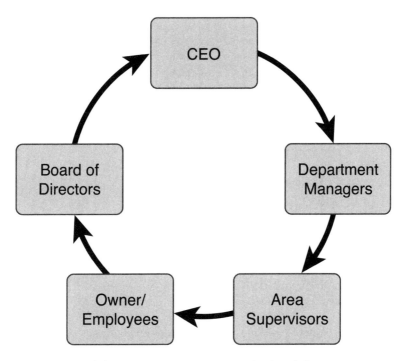

Exhibit 1 Kramer Sports organizational chart

13

UPS LOGISTICS AND TO MOVE TOWARD 4PL—OR NOT?

Remko van Hoek, Cranfield University, School of Management

Introduction

The fourth-party logistics provider (4PL) participates in supply chain coordination instead of just providing operational logistics and fulfillment services, like a traditional third-party logistics provider (3PL) would. The implementation of a 4PL model required a major transition for UPS, which has its heritage in express and physical logistics services, toward a more integral involvement in customers' supply chains, with a greater impact on customer supply chain performance and strategy. Sample UPS customer relations such as with Cisco Systems show how the potential of a 4PL approach for logistics service providers lies in areas such as increasing "value add" to customers, strengthening customer relations, and escaping the drive toward commoditization of logistics services. The question is, however, how feasible this model is in the longer term, especially when factoring in the perspective of the manufacturer and the existence of alternative candidates for outsourcing supply chain coordination, including consultants and contract manufacturers. Further, the question is how the migration from a 3PL to a 4PL concept impacts the transaction economics and relationship coordination requirements for both the provider and the client.

Industry Environment and Business Model Context

Logistics service providers have been keen on contributing to innovations in their client's supply chain for some time now and they have been expanding service offerings; for example, through the creation of 4PL offerings. The 4PL model essentially elevates the 3PL to a coordinator of the flow of goods, not just an operator in the physical movement of goods. This is seen by 3PLs as a method for not only increasing revenues, but also, more important, as a method to contribute to offering higher value-added activities in the supply chain than the warehousing and transport services traditionally offered. The market for these traditional services may still be growing; however, it is increasingly crowded with service providers offering cutthroat rates, an indicator of commoditization of the service. The 3PL model is also asset intensive, which in a price-sensitive market further challenges return on investments and financial performance. The 4PL model is far-less asset intensive as it focuses more on coordination, rather than just operating assets in service of the customer.[1]

Despite the availability of examples such as General Motors and Vector Logistics, in the market for expanded services, there are far fewer companies successfully operating. These services can upgrade the position of the 3PL in the supply chain beyond a supplier of commodity services, into a supplier of key services, such as coordination and management of the overall flow of goods in the supply chain, instead of merely operating the physical movement at selected links in the chain.

Exhibit 1 presents a service classification scheme from Brignall et al.[2] within which the evolution of 3PL services into the 4PL domain can be understood. What is really happening is that service providers are aiming to evolve from the mass services segment toward the professional services segment. This involves an increase in customization of services, contact time with customers (spent on developing customized service concepts, coordination efforts, and so on), as well as an increasing focus on customer relations and supply chain processes, instead of on equipment and products. As much sense as that might make from the 3PL perspective, the question is, will their customers really buy? In that respect, Coyle et al. state that

> While certain third party relationships do involve a very comprehensive set of service offerings, most customer-supplier relationships begin with a more modest set of activities to be managed by the third party. As customers grow accustomed to using the services of a third party for certain activities such as transportation and warehousing, they become candidates for a broader range of service offerings.[3]

Based upon a large-scale survey, van Hoek and van Dierdonck[4] concluded that, despite good efforts, 3PLs are not fully successful, if at all, in penetrating services such as customization and coordination that reach beyond their traditional areas of focus. Additionally, 3PLs are finding themselves facing a new set of competitors for the 4PL business. In addition to their traditional competition from transport companies and freight forwarders, consultants such as Accenture, software vendors, and contract manufacturers are also pursuing this business. One such contract manufacturer is the U.S.-based SCI; the company has split its logistics organization into two divisions, one focused solely on internal logistics and one focused on working with its customers on providing value-added outbound services. Adolfo Anzaldua, director of logistics operations, says the following:

> About a year and a half ago our external logistics requirements started taking off. We decided to break our logistics organization apart and create a dedicated group to providing the services our customers require, including direct order fulfillment, repair operations, reverse logistics, inventory hub management, and second stage manufacturing. As contract manufacturing outsourcing progression went from manufacturing to procurement and planning to full supply chain management and outsourcing, many 3PL partners were not able to fully service the OEMs and the contract manufacturing firms were quick to pick up on the services 3PLs could not handle.[5]

With this as general industry context, the next section introduces UPS's approach to expanding its service involvement in client relations using a 4PL model.

UPS Logistics Approach to 4PL

The business model of the 4PL as developed by UPS Logistics tends to evolve with customer relations; as displayed in Exhibit 2, it captures the evolution of client relations into 4PL accounts. Traditionally, various transport and warehousing companies and 3PLs have provided logistics services to original equipment manufacturers (OEMs) (phase A). Once a former 3PL develops a 4PL role, the supply structure changes to phase B in which the 4PL takes over the process and flow management of the OEM. The company takes over the logistics problems and starts managing the physical supply of logistics on behalf of the customer. In that step, an intermediate layer is created coordinating logistics service operations and providing the customer with a single point of contact. In the process, the service provider may occasionally take over customer employees with expertise about the product and/or client-customer relations in order to rapidly develop sufficient expertise.

The 4PL is often non-asset-focused and information-based. Those customer relations that have evolved from 3PL origins may have retained some hard assets, but the critical factor is a change of an inward-looking mentality from ownership of hard assets and asset utilization, to a more holistic focus of total supply chain effectiveness and process optimization utilizing existing 3PL capacity in the market. This means that the 4PL sources and coordinates operations on behalf of the customer. UPS Logistics may source services from UPS but will also source from other 3PLs depending upon which provider can offer the best value service (based upon service level, quality, consistency, and cost). The 3PL selection is sometimes partially based upon recommendations or stipulations from the client, but in those instances the involvement of that company is the responsibility of the client. UPS Logistics will involve them but if they perform poorly, then the client cannot hold UPS Logistics accountable, until UPS Logistics has endorsed that company after a period of performance evaluation.

In phase C, the 4PL begins to further develop a supply chain focus and starts to progress into a larger supply chain manager role for the customer. The 4PL begins to engage in supplier interfaces by calling off shipments of parts and components from the customer's supplies. This can be based upon call-off rules (once inventory goes under a certain level, order a replenishment) or can be based upon modeling. Furthermore, the 4PL starts to engage in customer-facing processes. The 4PL can receive and handle first- or second-level calls from the customer's ordering shipments and order fulfillment activities.

Once the leap to the supply chain level has been achieved, a second move is to engage in coordinating possible manufacturing interfaces, as shown in phase D. Knowing customer orders and supply operations, the 4PL can coordinate (final) manufacturing performed by contract manufacturers and end up in a position to leave the client with a virtual or design and marketing organization only, by integrating and running the entire supply chain for them. Obviously, there may be battles between the 4PL and the (contract) manufacturer on who is the best player to manage the supply chain, as there may be battles between 3PLs when progressing from phase A to B on who is the best player to perform the 4PL role. In this battle, implementation and management capabilities in the change process are key differentiators among players.

The resulting structure is one of a responsibility divide and a tiered supply chain. The client/manufacturer concentrates on supply chain strategy and coordination with customers and the 4PL. The 4PL coordinates with the manufacturer and logistics operators while managing logistics operations and the logistics operations focus on performing logistics activities and coordinating with the 4PL. This last point is possibly the most important responsibility for the manufacturer; outsourcing does not mean disposing of all responsibility at all. Ongoing coordination, key contacts, and system integration are among keys to sustainable success the manufacturer controls just as much, if not more, than the service provider.

Additionally, in order to enable the close working relationship between the 4PL and the manufacturer, certain levels of information technology (IT) systems integration will be required. At minimum, electronic order sharing and inventory reporting, but possibly also access to planning software and tools for the 4PL, will enhance its ability to operate as a seamlessly linked partner. In the case of UPS, the company uses its advance transportation network optimization tools and software to organize the customer's transportation; this can be a major benefit to the customer as it essentially directly taps into one of the key competitive capabilities and requirements for UPS.

Managing the Change Process

The process outlined in Exhibit 2 is one that reflects a disintegration and reintegration change process. Initially, the supply chain is disintegrated (phase A–D) with suppliers and different layers separated from the client. Reintegration, at a higher level of sophistication, is then achieved around the 4PL. How do you manage this transformation in the supply chain?

UPS Logistics does not participate in many "traditional" tender or contract bidding processes anymore. It generally consults to clients in supply chain design and reengineering projects. The company enters into a codesign process, which moves the relationship away from a sales and buying effort. UPS Logistics will bring logistical and supply chain experience to the table and codevelop the business model for a company in change or growth. A highly competitive market fee for consulting will be charged because of the anticipated supply chain management business opportunity for UPS Logistics. But if at the end of the process UPS Logistics is not commissioned as a 4PL, there is an additional charge to compensate for the effort. This places UPS Logistics in a more neutral position. Its initial focus is not on selling but on assisting the company in developing a business model utilizing existing and best-of-breed capacity in the marketplace while maintaining a sense of reality on the motives and motivation of those 3PLs and how to effectively daisy-chain them into a "seamless" collaborative physical supply chain for the client.

Once the 4PL model is in place (phase B), UPS Logistics begins to further interact with various functional areas of the client's organization. Manufacturing and marketing units, for example, have to be convinced of the prospect of the 4PL model; without this, the 4PL application remains more limited to transportation management (phase A). In developing the 4PL model into phases C and D, the company actually begins to run supply chain management for the customer's UPS functions as the flexible process glue in the supply chain operations once up and running. The change of mentality within the former 3PL and, more important, within the customer's organization, may be time consuming and demanding—for which benefits such as transparency and improved communication systems achieved in phase B are used as a stepping stone and a pitch with the other units of the customer's organization. This is different from solely using cost savings as the argument. This is the traditional argument used by 3PLs in the development of logistics services and has caused part of the commoditization in the industry. When the 4PL progresses into supply chain services, however, service enhancement becomes more critical, rather than just cost savings, signaling an escape from a more price-sensitive environment.

In that respect, the changes in operations, mentality, and supply chain management systems in progressing through the various stages of information integration show a relation with the stages in the contribution of logistics and supply chain to the competitiveness as identified by Bowersox and Closs.[6] They explain that logistics may initially contribute to cost savings. This is the initial pitch for the 4PL, very much like that for traditional 3PL. It can then progress into customer service enhancement, which is the advanced level of contribution from the 4PL, as explained. Finally, once synergies between the integrated flow of information and goods are actively managed new markets, products, and services can be created. This is also the final stage in the contribution of logistics to competitiveness, according to Bowersox and Closs.[7] Once the 4PL reaches phase D, it comes very close to this level. Through including manufacturing activities and adding supplementary services to the scope of logistics services, the client can achieve greater levels of customization for its customers. The logistics company can truly escape the commodity trap of traditional transport and warehousing services by offering expanding services with higher value added, such as described going up the vertical axis of Exhibit 1.

Changing Economics and Coordination

Once fully migrated to phase D, the economics of the relationship and transactions between the logistics service provider and the manufacturer have substantially changed. And as mentioned, the coordination between the two parties also has to change with the economics. Table 1 details typical changes moving from a 3PL to a 4PL relationship, whereas Table 3 depicts 4PL characteristics. It is important to note that a lot of these place responsibility on the manufacturer/client to contribute more extensively to the success of the relationship. These responsibilities include more extensive coordination, sharing of strategic changes and ambitions, more comprehensive performance measurement, and evaluation. All of these originate from the fact that the service provider integrates more deeply and comprehensively into the manufacturer's supply chain and the fact that the manufacturer becomes more dependent on the service provider for fulfillment performance to its customers.

Integration of IT systems is a particular dimension in the implementation of the 4PL concept that deserves consideration in the context of the economics and coordination of the relationship. A substantial amount of systems support will be necessary in the relationship, possibly including receiving of customer orders, ordering from suppliers electronically, and checking inventory levels in the manufacturing organization. This could be highly customer specific resulting in challenges with scale economics and speed of implementation and learning for the service provider. It is also likely that the manufacturer/client is going to demand that the service provider will establish the systems integration as part of its implementation efforts. It is likely, therefore, that the service provider will aim to use its existing systems such as UPS's customer desktop integrated tools for transportation and shipment management. UPS also has sophisticated network design systems it uses to help customers design supply chain networks and optimize its own transportation network, which can also be deployed.

Table 2 summarizes benefits for both the service provider and the manufacturer/client that can be accomplished when the migration is managed effectively.

Sample Client Relationship: Cisco Systems

A UPS Logistics "model" customer for the evolution into a 4PL is Cisco Systems. This relationship has evolved through phases A–D and currently involves both supplier- and customer-facing activities, on top of the coordination and management of all logistics flows in Europe, the Middle East, and North Africa.

UPS Logistics receives notification when products are ready for shipment from contract manufacturers and the Cisco plant. These products are then collected within 24 hours from one of more than 20 sites around the world, UPS books aircraft into the continent, and receives product into a dedicated 86,000-square-foot European logistics center, owned and operated by UPS.

On the outbound side, UPS selects the carrier and oversees delivery to customers throughout Europe, the Middle East, and North Africa. Using optimization software developed in-house by UPS, it consolidates shipments with a common destination. Orders frequently include components from multiple origins, creating the need for a system that can minimize the number of deliveries and reduce congestion at the loading dock.

For outbound shipments, UPS acts as a neutral party with respect to carrier choice. Various carrier algorithms (e.g., service level, price, and time in transit) have been populated in the system on a postal code level. The system provides a "mini RFQ" (request for quotation or proposal) to find the best carrier from an approved vendor list every time a new shipment is being presented for a customer. Throughout the process, the order status is communicated to Cisco so that it can provide its customer with continuous access to information. Until an order is fulfilled, a customer has the opportunity to make adjustments to the order, such as changes to delivery date. Every logistics movement registered in the UPS system is also registered immediately in the Cisco system. Annually, UPS Logistics handles more than one million boxes for Cisco.

UPS also runs an internal call center, which answers logistics-related questions by Cisco's Customer Service Department. Recently, UPS has begun pick-and-pack operations, combining accessories such as power cords with Cisco orders. Beyond that, says a logistics manager from Cisco, "The opportunities of the market will drive the relationship that we have with UPS."

Discussion Questions

Given that the development of the 4PL model is "up to the market," several questions face UPS and any 3PL considering or actively pursuing a 4PL strategy:

- Why would 3PLs consider a 4PL strategy, bearing in mind the nature of the 3PL industry?

- How are the economics of the relationship affected and what are the coordination requirements to ensure return for both parties?

- What criteria might (potential) customers use to evaluate possible 4PLs?

- How would 3PLs score on these criteria against competition (contract manufacturers, consultants, IT vendors, transport companies, and freight forwarders)? Or in other words, what are the critical success factors for a 4PL strategy?

- What conclusions can be drawn based upon that evaluation for the longer-term feasibility of the 4PL strategy for 3PLs? Additionally, what are policy implications for manufacturers/outsourcers?

Endnotes

1. See also: J. Bumstead and K. Cannons, "From 4PL to Managed Supply Chain Operations," *Logistics & Transport Focus* (May 2002).

2. T. J. Brignall et al. "Linking Performance Measures and Competitive Strategies in Service Business: Three Case Studies," in *Management Accounting Handbook* (Oxford: Butterworth Heinemann, 1997), 196–216.

3. John J. Coyle, Edward J. Bardi, and C. John Langley, *The Management of Business Logistics*, 5th edition (St. Paul: West Publishing Company, 1992), 550.

4. Remko I. van Hoek and Roland van Dierdonck, "Postponed Manufacturing Supplementary to Transportation Services," *Transportation Research Part E* 36 (2000): 205–217.

5. D. Hannon, "Contract Manufacturers Focus on Logistics Value Add," *Purchasing* (February 19, 2004).

6. Donald J. Bowersox and David Closs, *Logistical Management*, 5th edition (New York: McGraw-Hill, 1996).

7. Donald J. Bowersox and David Closs, *Logistical Management*, 5th edition (New York: McGraw-Hill, 1996).

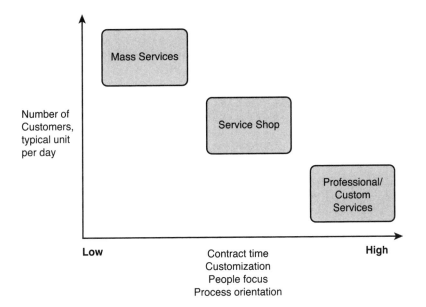

Exhibit 1   Service classification scheme

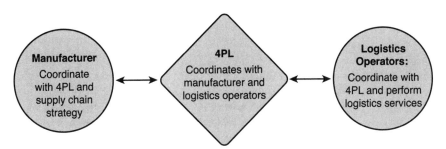

Exhibit 2   Responsibility divide and tiers in 4PL model

Table 1 Changing Economics and Coordination Moving from 3PL to 4PL

	3PL Relationship	4PL Relationship
Supply chain involvement of service provider	Physical movement and execution	Coordination and management of logistics operations
Asset intensity of services provided	High; trucks, warehouses, etc.	Lower; information systems, etc.
Knowledge intensity of services provided	Low; execution of standardized tasks	Higher; organization of flow of goods
Dependence of manfucturer on service execution for meeting customer demand	Medium; low switching costs and multiple similar providers of a commoditizing service	Higher; manufacturer has order fulfillment running on service provider's systems and relies on its ability to meet orders
Contact points in manufacturer organization	Execution-level, day-to-day contacts and management for contract negotiations	Preferably dedicated contract point and senior-level supply chain design and strategy coordination
Performance measurement of services provided	Can be limited to throughput and results measured related to payments and quarterly evaluations	More comprehensive measurement, including customer service and strategic supply chain measures
Strategic information sharing by manufacturer	Limited to informing 3PL about changes in logistics service levels, facilities, and other changes impacting logistical execution	More comprehensive, including customer and supplier lists, service policies, and priorities

Table 2 Benefits from 4PL Implementation for Manufacturer and 3PL

	Manufacturer/Client	Service Provider/3PL
Strategic considerations	Leveraging outsourcing benefits beyond execution tasks, including further lower internal head count (transportation planning, for example) and benefit from service provider scale	Escape commoditization of 3PL services and penetrate client supply chain more deeply and comprehensively as part of expansion of relationship beyond the transactional; increase client dependence

	Risks: increased dependence on service provider for meeting customer demand, specificity of transportation planning might lower benefits of service provider experience and scale	
Financial benefits	See above	Higher value-added services; beyond basic execution into management and coordination; lower asset-intensity
Operational benefits	See above	Expanded scope of operations serviced Risks: extensive systems development and integration and specificity of manufacturer's supply chain organization might limit ability to scale investment and lengthen learning process during the implementation stages

Table 3 4PL Evaluation Grid

	Transport Company	3PL (such as UPS)	Freight Forwarder	Management Consultant	ICT Provider	Contract Manufacturer
Focus within 4PL concept	A to B (physical movement)	Logistics/physical flow	Organizing for physical movement	**Concept development implementation**	ICT tools implementation	Good supply
Logistics capability	Limited to transport (possibly warehousing)	**High-speed delivery, logistics operational**	Transport organization for client	Possibly expertise from consulting, little hands-on experience	Possibly expertise from systems development	Limited, mostly inbound flows expertise
IT systems capability	Limited, possibly EDI	Increasingly prerequisite for operational logistics	Possibly EDI	Possibly expertise from consulting	**Core competence**	Possibly enterprise resource planning (ERP) for supply chain coordination
Information management capability	Focus on transport transaction; no supply coordination	Some logistics coordination prerequisite	Mostly in terms of organizing transport only	Part of problem solving, preparation, capability	**Part of system deployment capability**	Focus on manufacturing part of chain
Supply chain management capability	Very limited supply chain interface and involvement	Focus traditionally limited to logistics	Very limited supply chain interface	**Broad network in business and expertise**	Little experience in actual supply chain participation	Focus on manufacturing part of chain

	Change agent capability	Experience in implementing logistics concepts	Operational focus, limited client interface	Key competence is in change processes	Part of systems deployment capability	Experience in taking over large portions of value add in manufacturing
Reliance on own assets	Focus on transport capacity utilization	Focused on logistics capacity utilization	Little assets	People only asset	Only on IT tools and people	Focus on manufacturing asset utilization

INDEX

NUMERICS

3PL (third-party logistics provider), 70, 100, 116
   adoption of 4PL role, 175
   managing transition to 4PL, 177-178
   versus 4PL, 174-175

4PL (fourth-party logistics provider)
   business model context, 173-174
   UPS's adoption of
      Cisco Systems as client, 179
      integration of IT systems, 178
      managing the transformation from 3PL, 177-178
   UPS's approach to, 175-176
   UPS's implementation of, 173
   versus 3PL, 174-175

21st Century Book Chain Co., Ltd., 61
   distribution network, 69-70
   inbound logistics, 68
   network design, improving, 70-71
   suppliers, 64-65

99 Read, 62

Accenture, 174

Alliance for Children Everywhere case study
   allocation of foods to other aid organizations, 139-140
   goods turnover program, 140
   receipt of goods, 137
   routing of shipments, 134-136
   total logistics costs, 141, 143
   transportation documentation, 136
   transportation of goods, 138-139

Anzaldua, Adolfo, 175

assumptions for EOQ, 28

B

Bell, Mark, 87

benefits of case studies, 2

Bertelsmann AG, company history, 60

Bertelsmann Book Club case study, 59-60, 67-68
   business in China, 60
      Chinese book market, 61
      corporate culture, 61
   catalog development process, 63
   demand forecasting, 65-66
   distribution network, 69-70
   inbound logistics, 63
   network design, improving, 70-71
   outbound logistics, 64
   suppliers for Direct Group China, 62

Bertelsmann, Carl, 60

book market in China, 61

business model, Oklahoma Goodwill Industries, 46

C

Carnival Corporation case study, 82
   supplier recommendations, 93
   competitive benchmark data, 86
   food consumption variables, 84
   food operating policies, 85

food procurement process, 85
food supplier base, 86
North American cruise industry, 82
statement of operations, 89
supplier recommendations, 87-88

case studies
  benefits of, 2
  Bertelsmann Book Club, 59-60, 67-68
    business in China, 60-61
    catalog development process, 63
    Chinese book market, 61
    company history, 60
    demand forecast, 65-66
    distribution network, 69-70
    inbound logistics, 63
    network design, improving, 70-71
    outbound logistics, 64
    suppliers for Direct Group China, 62
  Carnival Corporation, 82
    competitive benchmark data, 86
    food consumption profile, 84
    food consumption variables, 84
    food operating policies, 85
    food procurement process, 85
    food supplier base, 86
    North American cruise industry, 82
    statement of operations, 89
    supplier recommendations, 87-88, 93
  Dockomo Heavy Machinery Equipment, Ltd., 5
    dead inventory, 8
    fast movers, monthly demand data for, 15
    first pick rate, 8
    forecasting, 9
    inventory list, 14
    medium movers, monthly demand data for, 17-18
    parts categorization, 13
    safety stock, 10
    slow movers, monthly demand data for, 19-20
    supply chain partners, 7-8
    very slow movers, monthly demand data for, 21-22
  DSM Manufacturing, 95
    company history, 96
    full truckload state-to-state cost per mile, 101, 104-105
    intermodal state-to-state costs, 104
    network analysis, 95
    ocean lane rates, 101-102
    Project Fragrance, 97
    transportation costs, minimizing, 98-101
  humanitarian logistics
    transportation of goods, 138-139
    allocation of foods to other aid organizations, 139-140
    goods turnover program, 140
    receipt of goods, 137
    routing of shipments, 134-136
    total logistics costs, 141-143
    transportation documentation, 136
  Innovative Distribution Company, 126
    domestic suppliers, 127
    global suppliers, 128
    new product sourcing, 127
    total cost of ownership, 126
  Kiwi Medical Devices, Ltd., 109-110
    company history, 110
    global sales, 110-111
    materials flow, 116-117
    offshoring, 113-114
    operating costs, 111
    shelter option, 115
    subcontracting, 115
    wholly owned subsidiaries option, 115

Kramer Sports, 165
- *company history, 165*
- *factory operations, 166-167*
- *lean production implementation, 168-170*

Megamart seasonal demand planning, 33
- *demand characteristics of gas grills, 34*
- *gas grill purchase plan, developing, 39-40*
- *past failures, 37-38*
- *physical characteristics of gas grills, 35*
- *stakeholder concerns, 35-37*

Oklahoma Goodwill Industries, 43
- *business model, 46*
- *company history, 45*
- *concerns for immediate future, 50, 53-54*
- *contract services, 46*
- *demand planning, 47-49*
- *logisitics management, 44-45, 49-50*
- *revenue sources, 46*
- *sales in secondary markets, 46*
- *supply uncertainty, 47, 49*

service purchasing, 149-150
- *category management, 151-152*
- *Gensin's sourcing guidelines, 151-152*
- *legal spending, 153-154*
- *purchasing reorganization, 150*

Silo Manufacturing Corporation, 27-28, 30
UPS, 4PL, 173-179

catalog development process, Bertelsmann Book Club, 63
category management, 151-152
charitable groups, Giving Hands, 134
China Science & Technology Book Company, 60
Chinese book market, 61
Chinese culture, impact on Western businesses, 61
Chou, Foguang, 59
Christian Alliance for Children in Zambia, 133
Cisco Systems, relationship with UPS, 179

Cole, Conrad, 153
Compact Rain, 100
company history, Oklahoma Goodwill Industries, 45
comparing 3PL and 4PL, 174-175
concerns for immediate future, Oklahoma Goodwill Industries case study, 50, 53-54
Cooper, Reginald, 81
Cortes, Gustavo, 97
Council of Supply Chain Management Professionals, 2
Craig, Timothy, 109

D

Davis, Chris, 43
Davis, Ed, 30
dead inventory, 8
demand characteristics of gas grills, 34
demand planning
- Oklahoma Goodwill Industries case study, 47-49
- Bertelsmann Book Club case study, 65-66

Direct Group China, suppliers, 62
Dockomo Heavy Machinery Equipment, Ltd., case study, 5
- dead inventory, 8
- fast movers, monthly demand data for, 15
- first pick rate, 8
- forecasting, 9
- inventory list, 14
- medium movers, monthly demand data for, 17-18
- parts categorization, 13
- safety stock, 10
- slow movers, monthly demand data for, 19-20
- supply chain partners, 7-8
- very slow movers, monthly demand data for, 21-22

donations as part of OGI's business model, 46

DSM Manufacturing case study, 95
    full truckload state-to-state cost per mile, 101, 104-105
    company history, 96
    intermodal state-to-state costs, 104
    network analysis, 95
    ocean lane rates, 101-102
    Project Fragrance, 97
    transportation costs, minimizing, 98-101

E

EatWell Distributors, 86
employee-owned cooperatives, Kramer Sports, 166
EOQ (economic order quantity), 28-30

F

factory operations, Kramer Sports, 166-167
fast movers, monthly demand data for (Dockomo Heavy Machinery Equipment, Ltd., case study), 15
Feng, Liu, 67
Ferguson, Fred, 30
first pick rate, 8
food supply chain, Carnival Corporation case study, 82
    competitive benchmark data, 86
    food consumption profile, 84
    food consumption variables, 84
    food operating policies, 85
    food procurement process, 85
    food supplier base, 86
    statement of operations, 89
    supplier recommendations, 87-88, 93
forecasting, Dockomo Heavy Machinery Equipment, Ltd., case study, 9

G

gas grills (Megamart case study)
    demand characteristics of, 34
    physical characteristics of, 35
    purchase plan, developing, 39-40
Gaylord, E. K., 45
Gensin case study, 149-150
    category management, 151-152
    legal spending, 153-154
    purchasing reorganization, 150
    sourcing guidelines, 151-152
Giordano, Luigi, 81
Giving Hands, 134-136
global sales, Kiwi Medical Devices, Ltd., 110-111

H

Harris, F. W., 28
Hazard, John L., 126
Helms, Rev. Edgar J., 45
Heskett, James L., 126
history
    of Bertelsmann AG, 60
    of DSM Manufacturing, 96
    of Kiwi Medical Devices, Ltd., 110
    of Kramer Sports, 165
House of Moses, 133
humanitarian logistics
    allocation of foods to other aid organizations, 139-140
    goods turnover program, 140
    receipt of goods, 137
    routing of shipments, 134-136
    total logistics costs, 141-143
    transportation documentation, 136
    transportation of goods, 138-139

I

IFF (international freight forwarder), 134
improving network design, Bertelsmann Book Club case study, 70-71
inbound logistics
    21st Century Book Chain Company, 68
    Bertelsmann Book Club, 63
Innovative Distribution Company case study
    domestic suppliers, 127
    global suppliers, 128
    new product sourcing, 127
    total cost of ownership, 126
intermodal state-to-state costs, 104
inventory, demand forecasting for Bertelsmann Book Club case study, 65-66
inventory list, Dockomo Heavy Machinery Equipment, Ltd., case study, 14

J-K

Kiwi Medical Devices, Ltd., case study, 109-110
    company history, 110
    global sales, 110-111
    materials flow, 116-117
    offshoring, 113-114
    operating costs, 111
    shelter option, 115
    subcontracting, 115
    wholly owned subsidiaries option, 115
Kohl, Dr. Helmut, 60
KPIs (key performance indicators), 65
Kramer Sports case study, 165
    company history, 165
    factory operations, 166-167
    lean production implementation, 168-170

L

lean production system (Kramer Sports case study), 165-167
    implementing, 168-169
    reasons for success, 170

Ledger, Michelle, 109
legal spending, services purchasing case study, 153-154
Lewin, Robert, 27
logistics
    3PL (UPS case study), 70, 100, 116
        adoption of 4PL role, 175
        managing transition to 4PL, 177-178
        versus 4PL, 174-175
    4PL, 173
        business model context, 173-174
        client relationships, 179
        integration of IT systems, 178
        managing transformation from 3PL, 177-178
        UPS's approach to, 175-176
    humanitarian logistics case study
        allocation of foods to other aid organizations, 139-140
        goods turnover program, 140-143
        receipt of goods, 137
        routing of shipments, 134-136
        transportation of goods, 138-139
    inbound logistics
        21st Century Book Chain Company, 68
        Bertelsmann Book Club case study, 63
    outbound logistics, Bertelsmann Book Club case study, 64
    service providers, 70
logistics management, Oklahoma Goodwill Industries case study, 44-45, 49-50

M

managing transformation from 3PL to 4PL, 177-178
Manrodtner, Steve, 34
Martin, Ferris, 27
materials flow, Kiwi Medical Devices, Ltd., case study, 116-117
May, Alice, 149
McConnell, Abe, 98

medium movers, monthly demand data for (Dockomo Heavy Machinery Equipment, Ltd., case study), 17-18

Megamart case study, 33
- gas grills
  - *demand characteristics, 34*
  - *failures to fix supply chain, 37-38*
  - *physical characteristics, 35*
  - *purchase plan, developing, 39-40*
- stakeholder concerns, 35-37

Mehra, Vinod, 5

minimizing transportation costs, 98-101
- full truckload state-to-state cost per mile, 101, 104-105
- intermodal state-to-state costs, 104
- ocean lane rates, 101-102

ModelPro 2020, 97

N

network analysis, 95-96

network design, improving (Bertelsmann Book Club case study), 70-71

North American cruise industry, 82

O

ocean lane rates, 101-102

OEMs (original equipment manufacturers), 175

offshoring (Kiwi Medical Devices, Ltd., case study), 109
- global sales, 110-111
- materials flow, 116-117
- operating costs, 111
- right-sourcing, 113-114
- shelter option, 115
- subcontracting, 115
- wholly owned subsidiaries option, 115

Oklahoma Goodwill Industries case study, 43
- business model, 46
- company history, 45
- concerns for immediate future, 50, 53-54

demand planning, 47-49

logistics management, 44-45, 49-50

revenue sources, 46

supply uncertainty, 47, 49

operating costs, Kiwi Medical Devices, Ltd., 111

outbound logistics, Bertelsmann Book Club, 64

P

parts categorization, Dockomo Heavy Machinery Equipment, Ltd., case study, 13

past failures to fix grill supply chain (Megamart case study), 37-38

Patrachalski, Peter, 27

physical characteristics of gas grills, 35

power distance, 61

Project Fragrance, 97

Q-R

revenue sources, Oklahoma Goodwill Industries, 46

RFIs (requests for information), 116

Rhodes, Julie, 97

right-sourcing, Kiwi Medical Devices, Ltd., 113
- materials flow, 116-117
- shelter option, 115
- subcontracting, 115
- wholly owned subsidiaries option, 115

Robertson, Heather, 43

Rousseau, Timothy, 83

S

safety stock, Dockomo Heavy Machinery Equipment, Ltd., case study, 10

SCM (supply chain management), 1
- 4PL
  - *business model context, 173-174*
  - *client relationships, 179*
  - *integration of IT systems, 178*

managing the transformation from 3PL, 177-178
UPS's approach to, 175-176
case studies, benefits of, 2
Innovative Distribution Company case study
domestic suppliers, 127
global suppliers, 128
total cost of ownership, 126-127
seasonal demand planning, Megamart case study, 33
    demand characteristics of gas grills, 34
    gas grill purchase plan, developing, 39-40
    past failures, 37-38
    physical characteristics of gas grills, 35
    stakeholder concerns, 35-37
services purchasing case study, 149-150
    category management, 151-152
    Gensin's sourcing guidelines, 151-152
    legal spending, 153-154
    purchasing reorganization, 150
Shanghai Bertelsmann Culture Industry Company, 60
shelter service providers, 115
Silo Manufacturing case study, 27-30
slow movers, monthly demand data for (Dockomo Heavy Machinery Equipment, Ltd., case study), 19-20
Smith, Steve, 30
sourcing guidelines for Gensin, 151-152
spare parts SCM, Dockomo Heavy Machinery Equipment, Ltd., case study, 5
    dead inventory, 8
    fast movers, monthly demand data for, 15
    first pick rate, 8
    forecasting, 9
    inventory list, 14
    medium movers, monthly demand data for, 17-18
    parts categorization, 13
    safety stock, 10
    slow movers, monthly demand data for, 19-20

supply chain partners, 7-8
very slow movers, monthly demand data for, 21-22
stakeholder concerns, Megamart case study, 35-37
suppliers for Direct Group China, 62
supply chain partners, Dockomo Heavy Machinery Equipment, Ltd., 7-8
supply uncertainty, Oklahoma Goodwill Industries case study, 47-49

T

Thielen, Gunther, 60
Thomas Consulting, 8
total logistics costs, humanitarian logistics case study, 141-143
transportation costs
    full truckload state-to-state cost per mile, 101, 104-105
    intermodal state-to-state costs, 104
    minimizing, 98-101
    ocean lane rates, 101-102

U-V

UPS case study, 4PL
    business model context, 173-174
    Cisco Systems as client, 179
    integration of IT systems, 178
    managing the transformation from 3PL, 177-178
    UPS's approach to, 175-176
Vaughn, John, 168
very slow movers, monthly demand data for (Dockomo Heavy Machinery Equipment, Ltd., case study), 21-22

W-X-Y-Z

Wilcox, Tim, 165
Wilson's EOQ, 28
World Vision, 140